AP* Physics 1 Essentials

An APlusPhysics Guide

2nd Edition

Dan Fullerton

**Physics Teacher
Irondequoit High School**

**Adjunct Professor
Microelectronic Engineering
Rochester Institute of Technology**

Dedication

This book is dedicated to the many students and educators
who have supported the APlusPhysics site and series of
instructional resources. Without your support and encouragement,
this project would have dissipated years ago. Thank you.

Credits

Thanks To:
Peter Geschke, Mike Powlin, Tom Schulte, Dolores Gende, Joe Kunz, Laurie Peslak,
Mike Dalessandro, Rob Spencer, Karen Finter, Scott Mathieson, Joanne Schwager,
Dan Burns, Amy Pezzoni, Bob Enck, Paul Sedita

Silly Beagle Productions
656 Maris Run
Webster, NY 14580
Internet: www.SillyBeagle.com
E-Mail: info@SillyBeagle.com

Cover Design:
Interior Illustrations by Dan Fullerton, Jupiterimages and NASA unless otherwise noted
All images and illustrations ©2014 Jupiterimages Corporation and Dan Fullerton
Edited by Joe Kunz

*AP, AP Physics 1, Advanced Placement Program, and College Board are registered trademarks of
the College Entrance Examination Board, which was not involved in the production of, and does
not endorse, this product.*

Sales and Ordering Information
http://www.aplusphysics.com/ap1
Sales@SillyBeagle.com
Volume discounts available.
E-book editions available.

Printed in the United States of America
ISBN: 978-0-9907243-0-8

1 2 3 4 5 6 7 8 9 0

Silly Beagle Productions

Welcome to <u>AP Physics 1 Essentials - an APlusPhysics Guide</u>. From mechanics to momentum and sound to circuits, this book is your essential physics resource for use as a standalone guide, a companion to classroom text, or as a review book for your AP Physics 1 course. What sets this book apart from other review books?

1. It reviews the essential concepts and understandings required for success in AP Physics 1, and is written specifically to prepare students for optimal performance on the AP Physics 1 Exam.
2. It includes more than 500 sample questions with full solutions, integrated into the chapters immediately following the material being covered, so you can test your understanding. You'll also find tons of AP-style problems at the end of the book.
3. It is supplemented by the free APlusPhysics.com website, which includes:
   a. Videos and tutorials on key physics concepts
   b. Interactive practice quizzes
   c. Discussion and homework help forums supported by the author and fellow readers
   d. Student blogs to share challenges, successes, hints and tricks
   e. Projects and activities designed to improve your understanding of essential physics concepts in a fun and engaging manner
   f. Latest and greatest physics news

Just remember, physics is fun! It's an exciting course, and with a little preparation and this book, you can transform your quest for essential physics comprehension from a stressful chore into an enjoyable and, yes, FUN, opportunity for success.

How to Use This Book

This book is arranged by topic, with sample problems and solutions integrated right in the text. Actively explore each chapter. Cover up the in-text solutions with an index card, get out a pencil, and try to solve the sample problems yourself as you go, before looking at the answer. If you're stuck, don't stress! Post your problem on the APlusPhysics website (http://aplusphysics.com) and get help from other students, teachers, and subject matter experts (including the author of this book!). Once you feel confident with the subject matter, test yourself and see how you perform. Review areas of difficulty, then try again and watch your understanding improve!

* There are a number of topics presented in this book which are typically included in an introductory physics course but are not specifically tested on the AP Physics 1 Exam. These topics are marked with an asterisk to allow those focused solely on preparing for the AP Physics 1 Exam to bypass them in favor of topics that they know will be tested.

Table of Contents

~ Day 1

~ Phy 2

Chapter 1: Introduction

"There is a theory which states that if ever anybody discovers exactly what the Universe is for and why it is here, it will instantly disappear and be replaced by something even more bizarre and inexplicable.

There is another theory which states that this has already happened."

— Douglas Adams

Objectives

Explore the scope of AP Physics 1 and review prerequisite skills for success.

1. Recognize the questions of physics.
2. List several disciplines within the study of physics.
3. Describe the key topics covered in AP Physics 1.
4. Define matter, mass, work and energy.
5. Explain how systems are comprised of constituent substructures.
6. Recognize when a system may be treated as an object.

What is Physics?

Physics is many things to many different people. If you look up physics in the dictionary, you'll probably learn physics has to do with matter, energy, and their interactions. But what does that really mean? What is matter? What is energy? How do they interact? And most importantly, why do we care?

Physics, in many ways, is the answer to the favorite question of most 2-year-olds: "Why?" What comes after the why really doesn't matter. If it's a "why" question, chances are it's answered by physics: Why is the sky blue? Why does the wind blow? Why do we fall down? Why does my teacher smell funny? Why do airplanes fly? Why do the stars shine? Why do I have to eat my vegetables? The answer to all these questions, and many more, ultimately reside in the realm of physics.

Matter, Systems, and Mass

If physics is the study of matter, then we probably ought to define matter. **Matter**, in scientific terms, is anything that has mass and takes up space. In short, then, matter is anything you can touch – from objects smaller than electrons to stars hundreds of times larger than the sun. From this perspective, physics is the mother of all science. From astronomy to zoology, all other branches of science are subsets of physics, or specializations inside the larger discipline of physics.

In physics, you'll often times talk about **systems** as collections of smaller constituent substructures such as atoms and molecules. The properties and interactions of these substructures determine the properties of the system. In many cases, when the properties of the constituent substructures are not important in understanding the behavior of the system as a whole, the system is referred to as an **object**. For example, a baseball is comprised of many atoms and molecules which determine its properties. For the purpose of analyzing the path of a thrown ball in the air, however, the makeup of its constituent substructures is not important; therefore, we treat the ball as a single object.

So then, what is mass? **Mass** is, in simple terms, the amount of "stuff" an object is made up of. But of course, there's more to the story. Mass is split into two types: **inertial mass** and **gravitational mass**. Inertial mass is an object's resistance to being accelerated by a force. More massive objects accelerate less than smaller objects given an identical force. Gravitational mass, on the other hand, relates to the amount of gravitational force experienced by an object. Objects with larger gravitational mass experience a larger gravitational force.

Confusing? Don't worry! As it turns out, in all practicality, inertial mass and gravitational mass have always been equal for any object measured, even if it's not immediately obvious why this is the case (although with an advanced study of Einstein's Theory of General Relativity you can predict this outcome).

1.1 Q: On the surface of Earth, a spacecraft has a mass of 2.00×10^4 kg. What is the mass of the spacecraft at a distance of one Earth radius above Earth's surface?

(A) 5.00×10^3 kg

(B) 2.00×10^4 kg

(C) 4.90×10^4 kg

(D) 1.96×10^5 kg

1.1 A: (B) 2.00×10^4 kg.   Mass is constant, therefore the spacecraft's mass at a distance of one Earth radius above Earth's surface is 2.00×10^4 kg.

Energy

If it's not matter, what's left?   Why, energy, of course.   As energy is such an everyday term that encompasses so many areas, an accurate definition can be quite elusive.   Physics texts oftentimes define **energy** as the ability or capacity to do work.   It's a nice, succinct definition, but leads to another question – what is work?   **Work** can also be defined many ways, but a general definition starts with the process of moving an object.   If you put these two definitions together, you can vaguely define energy as the ability or capacity to move an object.

Mass – Energy Equivalence

So far, the definition of physics boils down to the study of matter, energy, and their interactions. Around the turn of the 20th century, however, several physicists began proposing a strong relationship between matter and energy. Albert Einstein, in 1905, formalized this with his famous formula $E=mc^2$, which states that the mass of an object, a key characteristic of matter, is really a measure of its energy. This discovery has paved the way for tremendous innovation ranging from nuclear power plants to atomic weapons to particle colliders performing research on the origins of the universe. Ultimately, if traced back to its origin, the source of all energy on earth is the conversion of mass to energy!

Scope of AP Physics 1

Physics, in some sense, can therefore by defined as the study of just about everything. Try to think of something that isn't physics – go on, I dare you! Not so easy, is it? Even the more ambiguous topics can be categorized as physics. A Shakespearean sonnet? A sonnet is typically read from a manuscript (matter), sensed by the conversion of light (energy) alternately reflected and absorbed from a substrate, focused by a lens in the eye, and converted to chemical and electrical signals by photoreceptors on the retina. It is then transferred as electrical and chemical signals along the neural pathways to the brain. In short, just about everything is physics from a certain perspective.

As this is an introductory book in physics, it's important to limit the scope of study to a foundational understanding. You'll begin with a study of objects in motion, also known as kinematics, and then explore how the motion of an object is changed by application of a force, known as dynamics. The motion of an object can also be changed by a collision or explosion, studied in the linear momentum chapter. Next, you'll explore objects moving in circular paths, including the force that causes orbital motion, gravity. Your study of rotational motion will continue as you explore rotational kinematics and rotational dynamics in the rotational motion chapter. You'll then learn about work, energy, and power before concluding the mechanics section of the book by learning about oscillations and simple harmonic motion.

The next section of the book focuses on mechanical waves and sound. You will learn about wave characteristics, sound waves, interference, superposition, and even the Doppler Effect.

From there, you'll dive into basic electrostatics, learning about the building blocks of matter, electrical charge, electrical forces, electrical fields, and electric potential difference (a.k.a. voltage). Finally, the AP Physics 1 portion of the course will conclude with a study of basic circuits, beginning with current, resistance, and Ohm's Law, and finishing with basic DC circuit analysis.

The AP Physics 1 Exam itself is a three-hour exam designed to test your understanding of physics through a series of problems which require content knowledge, scientific inquiry skills, and the ability to reason logically and think critically. This is not an exam you can cram for, as it tests process skills such as your ability to apply physics concepts in unfamiliar contexts, explain physical relationships, design experiments, analyze data, and make generalizations from multiple areas of study. Instead, you should focus on building a strong understanding of the underlying principles emphasized in your course and in this book. To assist in this endeavor, the College Board has even published a list of seven "Big Ideas" that are prevalent throughout the study of physics, as well as associated "Enduring Understandings" that support these big ideas. These ideas and understandings will be emphasized throughout each chapter and highlighted in the chapter objectives.

7 Big Ideas in Physics

1. Objects and systems have properties such as mass and charge. Systems may have internal structure.
2. Fields existing in space can be used to explain interactions.
3. The interactions of an object with other objects can be described by forces.
4. Interactions between systems can result in changes in those systems.
5. Changes that occur as a result of interactions are constrained by conservation laws.
6. Waves can transfer energy and momentum from one location to another without the permanent transfer of mass and serve as a mathematical model for the description of other phenomena.
7. The mathematics of probability can be used to describe the behavior of complex systems and to interpret the behavior of quantum mechanical systems.

from the "AP Physics 1 & AP Physics 2 Curriculum Framework, 2014-2015"

In addition, the AP Physics course requires students to develop proficiency in seven "science practices" which are crucial to success. These practices, or skills, are best developed through an ongoing series of hands-on explorations and laboratory activities. As important as content knowledge is, physics is something you do, not just something you know. Therefore the lab component is an important part of any AP Physics course.

7 Science Practices

1. Use representations and models to communicate and solve scientific problems.
2. Use mathematics appropriately.
3. Engage in scientific questioning to extend thinking or guide investigations.
4. Plan an experiment and collect data to answer a scientific question.
5. Analyze data and evaluate evidence.
6. Work with scientific explanations and theories.
7. Connect, relate, and apply knowledge across various concepts and models.

adapted from the "AP Physics 1 & AP Physics 2 Curriculum Framework, 2014-2015"

The test itself consists of two sections: a 90-minute multiple choice section and a 90-minute free response section. The multiple choice section consists of 50 to 55 questions with four answer choices per question. Unlike most multiple choice tests, however, certain questions may have multiple correct items that need to be chosen to receive full credit. The free response section consists of five questions. Typically one question covers experimental design, one question covers quantitative and qualitative problem solving and reasoning, and three questions are of the short answer variety.

Note that the in-chapter questions are designed to assist you in solidifying the fundamental, underlying concepts required to succeed in the class. Most of these questions, therefore, are not at the level of complexity and integration across topics that you will see on the exam itself. They are designed to allow you to break the course into small, digestible chunks which you can then use to greater efficacy in labs, discussions, explorations, and deeper-understanding problems.

In preparing for the actual AP-1 exam itself, however, it is highly recommended that you practice a significant number of AP-1 style problems. More than 90 of these types of problems are available at the end of the book (with answers) in the appendix. Use these problems to test your understanding and readiness for the exam itself. You can find additional AP-1 style problems from the APlusPhysics website as well as the College Board's AP Physics 1 website.

Chapter 2: Math Review

"Mathematics is the door and key to the sciences."
— *Roger Bacon*

"Do not worry about your difficulties in mathematics. I assure you that mine are greater."
— *Albert Einstein*

Objectives

Review prerequisite math skills necessary for success.

1. Express answers correctly with respect to significant figures.
2. Use scientific notation to express physical values efficiently.
3. Convert and estimate SI units.
4. Differentiate between scalar and vector quantities.
5. Use scaled diagrams to represent and manipulate vectors.
6. Determine x- and y-components of two-dimensional vectors.
7. Determine the angle of a vector given its components.

Although physics and mathematics are not the same thing, they are in many ways closely related. Just like English is the language of this content, mathematics is the language of physics. A solid understanding of a few simple math concepts will allow you to communicate and describe the physical world both efficiently and accurately.

Significant Figures

Significant Figures (or sig figs, for short) represent a manner of showing which digits in a number are known to some level of certainty. But how do you know which digits are significant? There are some rules to help with this. If you start with a number in scientific notation:

- All non-zero digits are significant.
- All digits between non-zero digits are significant.
- Zeroes to the left of significant digits are not significant.
- Zeroes to the right of significant digits are significant.

When you make a measurement in physics, you want to write what you measured using significant figures. To do this, write down as many digits as you are absolutely certain of, then take a shot at one more digit as accurately as you can. These are your significant figures.

2.1 Q: How many significant figures are in the value 43.74 km?
2.2 A: 4 (four non-zero digits)

2.2 Q: How many significant figures are in the value 4302.5 g?
2.2 A: 5 (All non-zero digits are significant and digits between non-zero digits are significant.)

2.3 Q: How many significant figures are in the value 0.0083s?
2.3 A: 2 (All non-zero digits are significant. Zeroes to the left of significant digits are not significant.)

2.4 Q: How many significant figures are in the value 1.200×10^3 kg?
2.4 A: 4 (Zeroes to the right of significant digits are significant.)

As the focus of this book is building a solid understanding of basic physics concepts and applications, significant figures will not be emphasized in routine problem solving, but realize that in certain environments they can be of the highest importance. For the purposes of the AP Physics 1 exam, typically 3-4 significant figures will be adequate.

Scientific Notation

Because measurements of the physical world vary so tremendously in size (imagine trying to describe the distance across the United States in units of hair thicknesses), physicists oftentimes use what is known as **scientific notation** to represent very large and very small numbers. These very large and very small numbers would become quite cumbersome to write out repeatedly. Imagine writing 4,000,000,000,000 over and over again. Your hand would get tired and your pen would rapidly run out of ink! Instead, it's much easier to write this number as 4×10^{12}. Or on the smaller scale, the thickness of the insulating layer (known as a gate dielectric) in the integrated circuits that power computers and other electronics can be less than 0.000000001 m. It's easy to lose track of how many zeros you have to deal with, so scientists instead would write this number as 1×10^{-9} m. See how much simpler life can be with scientific notation?

Scientific notation follows a few simple rules. Start by showing all the significant figures in the number you're describing, with the decimal point after the first significant digit. Then, show your number being multiplied by 10 to the appropriate power in order to give you the correct value.

It sounds more complicated than it is. Let's say, for instance, you want to show the number 300,000,000 in scientific notation (a very useful number in physics), and let's assume you know this value to three significant digits. You would start by writing the three significant digits, with the decimal point after the first digit, as "3.00". Now, you need to multiply this number by 10 to some power in order to get back to the original value. In this case, you multiply 3.00 by 10^8, for an answer of 3.00×10^8. Interestingly, the power you raise the 10 to is exactly equal to the number of digits you moved the decimal to the left as you converted from standard to scientific notation. Similarly, if you start in scientific notation, to convert to standard notation, all you have to do is remove the 10^8 power by moving the decimal point eight digits to the right. Presto, you're an expert in scientific notation!

But, what do you do if the number is much smaller than one? Same basic idea. Let's assume you're dealing with the approximate radius of an electron, which is 0.00000000000000282 m. It's easy to see how unwieldy this could become. You can write this in scientific notation by writing out three significant digits, with the decimal point after the first digit, as "2.82." Again, you multiply this number by some power of 10 in order to get back to the original value. Because your value is less than 1, you need to use negative powers of 10. If you raise 10 to the power -15, specifically, you get a final value of 2.82×10^{-15} m. In essence, for every digit you moved the decimal place, you add another power of 10. And if you start with scientific notation, all you do is move the decimal place left one digit for every negative power of 10.

2.5 Q: Express the number 0.000470 in scientific notation.

2.5 A: 4.70×10^{-4}

2.6 Q: Express the number 2,870,000 in scientific notation.

2.6 A: 2.87×10^6

2.7 Q: Expand the number 9.56×10^{-3}.

2.7 A: 0.00956

2.8 Q: Expand the number 1.11×10^{7}.

2.8 A: 11,100,000

Metric System

Physics involves the study, prediction, and analysis of real-world phenomena. To communicate data accurately, you must set specific standards for basic measurements. The physics community has standardized what is known as the **Système International** (SI), which defines seven baseline measurements and their standard units, forming the foundation of what is called the metric system of measurement. The SI system is oftentimes referred to as the **mks system**, as the three most common measurement units are meters, kilograms, and seconds, which will be the focus for the majority of this course. The fourth SI base unit you'll use in this course, the ampere, will be introduced in the current electricity section.

The base unit of length in the metric system, the meter, is roughly equivalent to the English yard. For smaller measurements, the meter is divided up into 100 parts, known as centimeters, and each centimeter is made up of 10 millimeters. For larger measurements, the meter is grouped into larger units of 1000 meters, known as a kilometer. The length of a baseball bat is approximately one meter, the radius of a U.S. quarter is approximately a centimeter, and the diameter of the metal in a wire paperclip is roughly one millimeter.

The base unit of mass, the kilogram, is roughly equivalent to two U.S. pounds. A cube of water 10 cm x 10 cm x 10 cm has a mass of 1 kilogram. Kilograms can also be broken up into larger and smaller units, with commonly used measurements of grams (1/1000th of a kilogram) and milligrams (1/1000th of a gram). The mass of a textbook is approximately 2 to 3 kilograms, the mass of a baseball is approximately 145 grams, and the mass of a mosquito is 1 to 2 milligrams.

The base unit of time, the second, is likely already familiar. Time can also be broken up into smaller units such as milliseconds (10^{-3} seconds), microseconds (10^{-6} seconds), and nanoseconds (10^{-9} seconds), or grouped into larger units such as minutes (60 seconds), hours (60 minutes), days (24 hours), and years (365.25 days).

The metric system is based on powers of 10, allowing for easy conversion from one unit to another. A chart showing the meaning of commonly used metric prefixes and their notations can be extremely valuable in performing unit conversions.

Prefixes for Powers of 10		
Prefix	Symbol	Notation
tera	T	10^{12}
giga	G	10^9
mega	M	10^6
kilo	k	10^3
deci	d	10^{-1}
centi	c	10^{-2}
milli	m	10^{-3}
micro	μ	10^{-6}
nano	n	10^{-9}
pico	p	10^{-12}

Converting from one unit to another can be easily accomplished if you use the following procedure.

1. Write your initial measurement with units as a fraction over 1.
2. Multiply your initial fraction by a second fraction, with a numerator (top number) having the units you want to convert to, and the denominator (bottom number) having the units of your initial measurement.
3. For any units on the top right-hand side with a prefix, determine the value for that prefix. Write that prefix in the right-hand denominator. If there is no prefix, use 1.
4. For any units on the right-hand denominator with a prefix, write the value for that prefix in the right-hand numerator. If there is no prefix, use 1.
5. Multiply through the problem, taking care to accurately record units. You should be left with a final answer in the desired units.

Let's take a look at a sample unit conversion:

2.9 Q: Convert 23 millimeters (mm) to meters (m).

2.9 A: Step 1. $\dfrac{23mm}{1}$

        Step 2. $\dfrac{23mm}{1} \times \dfrac{m}{mm}$

        Step 3. $\dfrac{23mm}{1} \times \dfrac{m}{1mm}$

        Step 4. $\dfrac{23mm}{1} \times \dfrac{10^{-3}\,m}{1mm}$

        Step 5. $\dfrac{23mm}{1} \times \dfrac{10^{-3}\,m}{1mm} = 2.3 \times 10^{-2}\,m$

Now, try some on your own!

2.10 Q: Convert 2.67×10^{-4} m to mm.

2.10 A: $\dfrac{2.67 \times 10^{-4}\,m}{1} \times \dfrac{1mm}{10^{-3}\,m} = 0.267\,mm$

2.11 Q: Convert 14 kg to mg.

2.11 A: $\dfrac{14kg}{1} \times \dfrac{10^{3}\,mg}{10^{-3}\,kg} = 14 \times 10^{6}\,mg$

2.12 Q: Convert 3,470,000 µs to s.

2.12 A: $\dfrac{3,470,000\mu s}{1} \times \dfrac{10^{-6}\,s}{1\mu s} = 3.47s$

Accuracy and Precision

When making measurements of physical quantities, how close the measurement is to the actual value is known as the **accuracy** of the measurement. **Precision**, on the other hand, is the repeatability of a measurement. A common analogy involves an archer shooting arrows at the target. The bullseye of the target represents the actual value of the measurement.

Low Accuracy
Low Precision

High Accuracy
Low Precision

Low Accuracy
High Precision

High Accuracy
High Precision

Ideally, measurements in physics should be both accurate and precise.

Algebra and Trigonometry

Just as you find the English language a convenient tool for conveying your thoughts to others, you need a convenient language for conveying your understanding of the world around you in order to understand its behavior. The language most commonly (and conveniently) used to describe the natural world is mathematics. Therefore, to understand physics, you need to be fluent in the mathematics of the topics you'll study in this book... specifically basic algebra and trigonometry.

Now don't you fret or frown. You need only the most basic of algebra and trigonometry in order to successfully solve a wide range of physics problems.

A vast majority of problems requiring algebra can be solved using the same problem solving strategy. First, analyze the problem and write down what you know, what you need to find, and make a picture or diagram to better understand the problem if it makes sense. Then, start your solution by searching for a path that will lead you from your givens to your finds. Once you've determined an appropriate pathway (and there may be more than one), solve your problem algebraically for your answer. Finally, as your last steps, substitute in any values with units into your final equation, and solve for your answer, with units.

The use of trigonometry, the study of right triangles, can be distilled down to the definitions of the three basic trigonometric functions. When you know the length of two sides of a right triangle, or the length of one side and a non-right angle, you can solve for all the angles and sides of the triangle. If you can use the definitions of the sine, cosine, and tangent, you'll be fine in physics.

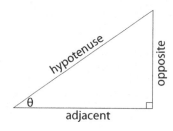

$$\sin\theta = \frac{opposite}{hypotenuse}$$

$$\cos\theta = \frac{adjacent}{hypotenuse}$$

$$\tan\theta = \frac{opposite}{adjacent}$$

Of course, if you need to solve for the angles themselves, you can use the inverse trigonometric functions.

$$\theta = \sin^{-1}\left(\frac{opposite}{hypotenuse}\right) = \cos^{-1}\left(\frac{adjacent}{hypotenuse}\right) = \tan^{-1}\left(\frac{opposite}{adjacent}\right)$$

2.13 Q: A car travels from the airport 14 miles east and 7 miles north to its destination. What direction should a helicopter fly from the airport to reach the same destination, traveling in a straight line?

2.13 A:

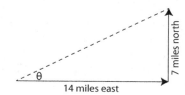

$$\theta = \tan^{-1}\left(\frac{opposite}{adjacent}\right)$$

$$\theta = \tan^{-1}\left(\frac{7\ miles}{14\ miles}\right) = 26.6°$$

2.14 Q: The sun creates a 10-meter-long shadow across the ground by striking a flagpole when the sun is 37 degrees above the horizon. How tall is the flagpole?

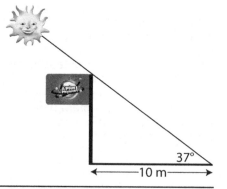

2.14 A: First, create a diagram of the situation, then recognize the shadow length is the adjacent side of the triangle, and the flagpole itself forms the opposite side. You can then use the tangent function to solve for the height of the flagpole.

$$\tan\theta = \frac{opp}{adj} \rightarrow opp = adj \times \tan\theta = (10m)\tan(37°) = 7.5m$$

Vectors and Scalars

Quantities in physics are used to represent real-world measurements, and therefore physicists use these quantities as tools to better understand the world. In examining these quantities, there are times when just a number, with a unit, can completely describe a situation. These numbers, which have just a **magnitude**, or size, are known as **scalars**. Examples of scalars include quantities such as temperature, mass, and time. At other times, a quantity is more descriptive if it also includes a direction. These quantities which have both a magnitude and direction are known as **vectors**. Vector quantities you may be familiar with include force, velocity, and acceleration.

Most students will be familiar with scalars, but to many, vectors may be a new and confusing concept. By learning just a few rules for dealing with vectors, you'll find that they can be a powerful tool for problem solving.

Vectors are often represented as arrows, with the length of the arrow indicating the magnitude of the quantity, and the direction of the arrow indicating the direction of the vector. In the figure below, vector B has a magnitude greater than that of vector A even though vectors A and B point in the same direction. It's also important to know that vectors can be moved anywhere in space. The positions of A and B could be reversed, and the individual vectors would retain their values of magnitude and direction.

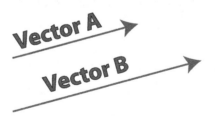

To add vectors A and B below, all you have to do is line them up so that the tip of the first vector touches the tail of the second vector. Then, to find the sum of the vectors, known as the **resultant**, draw a straight line from the start of the first vector to the end of the last vector. This method works with any number of vectors.

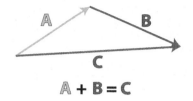

$$A + B = C$$

So how do you subtract two vectors? Try subtracting B from A. You can start by rewriting the expression A - B as A + -B. Now it becomes an addition problem. You just have to figure out how to express –B. This is easier than it sounds. To find the opposite of a vector, just point the vector in the opposite direction. Therefore, you can use what we already know about the addition of vectors to find the resultant of A-B.

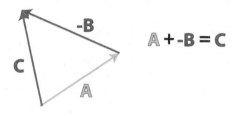

$$A + -B = C$$

Components of Vectors

You'll learn more about vectors as you go, but before moving on, there are a few basic skills to master. Vectors at angles can be challenging to deal with. By transforming a vector at an angle into two vectors, one parallel to the x-axis and one parallel to the y-axis, you can greatly simplify problem solving. To break a vector up into its components, you can use the basic trig functions.

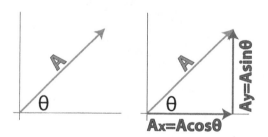

2.15 Q: The vector diagram below represents the horizontal component, F_H, and the vertical component, F_V, of a 24-newton force acting at 35° above the horizontal. What are the magnitudes of the horizontal and vertical components?

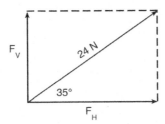

(A) F_H=3.5 N and F_V=4.9 N

(B) F_H=4.9 N and F_V=3.5 N

(C) F_H=14 N and F_V=20 N

(D) F_H=20 N and F_V=14 N

2.15 A: (D) $F_H = A_x = A\cos\theta = (24N)\cos 35° = 20N$

$$F_V = A_y = A\sin\theta = (24N)\sin 35° = 14N$$

2.16 Q: An airplane flies with a velocity of 750 kilometers per hour, 30° south of east. What is the magnitude of the plane's eastward velocity?

(A) 866 km/h

(B) 650 km/h

(C) 433 km/h

(D) 375 km/h

2.16 A:

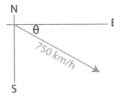

(B) $v_x = v\cos\theta = (750\,^{km}/_h)\cos(30°) = 650\,^{km}/_h$

2.17 Q: A soccer player kicks a ball with an initial velocity of 10 m/s at an angle of 30° above the horizontal. The magnitude of the horizontal component of the ball's velocity is

(A) 5.0 m/s

(B) 8.7 m/s

(C) 9.8 m/s

(D) 10 m/s

2.17 A:

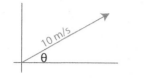

(B) $v_x = v \cos\theta = (10\,\text{m/s})\cos(30°) = 8.7\,\text{m/s}$

2.18 Q: A child kicks a ball with an initial velocity of 8.5 meters per second at an angle of 35° with the horizontal, as shown. The ball has an initial vertical velocity of 4.9 meters per second. The horizontal component of the ball's initial velocity is approximately

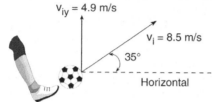

(A) 3.6 m/s
(B) 4.9 m/s
(C) 7.0 m/s
(D) 13 m/s

2.18 A: (C) $v_x = v \cos\theta = (8.5\,\text{m/s})\cos(35°) = 6.96\,\text{m/s}$

In similar fashion, you can use the components of a vector in order to build the original vector. Graphically, if you line up the component vectors tip-to-tail, the original vector runs from the starting point of the first vector to the ending point of the last vector. To determine the magnitude of the resulting vector algebraically, just apply the Pythagorean Theorem.

2.19 Q: A motorboat, which has a speed of 5.0 meters per second in still water, is headed east as it crosses a river flowing south at 3.3 meters per second. What is the magnitude of the boat's resultant velocity with respect to the starting point?
(A) 3.3 m/s
(B) 5.0 m/s
(C) 6.0 m/s
(D) 8.3 m/s

2.19 A: (C) 6.0 m/s

The motorboat's resultant velocity is the vector sum of the motorboat's speed and the riverboat's speed.

$$a^2 + b^2 = c^2$$
$$c = \sqrt{a^2 + b^2}$$
$$c = \sqrt{(5\,{}^m/_s)^2 + (3.3\,{}^m/_s)^2}$$
$$c = 6\,{}^m/_s$$

2.20 Q: A dog walks 8.0 meters due north and then 6.0 meters due east. Determine the magnitude of the dog's total displacement.

2.20 A: $$a^2 + b^2 = c^2$$
$$c = \sqrt{a^2 + b^2}$$
$$c = \sqrt{(6m)^2 + (8m)^2}$$
$$c = 10m$$

2.21 Q: A 5.0-newton force could have perpendicular components of

(A) 1.0 N and 4.0 N

(B) 2.0 N and 3.0 N

(C) 3.0 N and 4.0 N

(D) 5.0 N and 5.0 N

2.21 A: (C) The only answers that fit the Pythagorean Theorem are 3.0 N and 4.0 N ($3^2 + 4^2 = 5^2$)

2.22 Q: A vector makes an angle, θ, with the horizontal. The horizontal and vertical components of the vector will be equal in magnitude if angle θ is

(A) 30°

(B) 45°

(C) 60°

(D) 90°

2.22 A: (B) 45°. $A_x = A\cos(\theta)$ will be equal to $A_y = A\sin(\theta)$ when angle θ=45° since cos(45°)=sin(45°).

Vectors can also be added algebraically by adding their components to find the resultant vector.

2.23 Q: Kerbin the mouse travels 3 meters at an angle of 30 degrees north of east. He then travels 2 meters directly north. Finally, he travels 2 meters at an angle of 60 degrees south of west. What is the final position of Kerbin compared to his starting point?

2.23 A: You can first diagram this motion graphically, labeling the three different portions of the mouse's travel as vectors **A**, **B**, and **C**. By adding these three vectors to get resultant vector **R**, you determine Kerbin's final position. Start by lining up the three vectors tip-to-tail, drawing them to scale and using a protractor to make exact angles, then draw a vector from the starting point of the first vector to the ending point of the last vector to determine the resultant vector **R**.

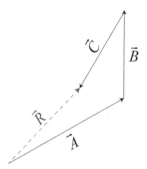

Alternately, you could break vectors **A**, **B**, and **C** into their components, then add up the individual x- and y-components to find the x- and y-components of resultant vector **R**.

$$\vec{A} = < A\cos\theta, A\sin\theta > = < 3m\cos 30°, 3m\sin 30° > = < 2.60m, 1.5m >$$
$$\vec{B} = < B\cos\theta, B\sin\theta > = < 2m\cos 90°, 2m\sin 90° > = < 0m, 2m >$$
$$\vec{C} = < C\cos\theta, C\sin\theta > = < -2m\cos 60°, -2m\sin 60° > = < -1m, -1.73m >$$

The resultant vector **R** is then found by adding up the x-components and y-components of the constituent vectors.

$$R_X = A_X + B_X + C_X = 2.60m + 0m + -1m = 1.6m$$
$$R_Y = A_Y + B_Y + C_Y = 1.5m + 2m + -1.73m = 1.77m$$
$$\vec{R} = < R_X, R_Y > = < 1.6m, 1.77m >$$

Therefore, the resultant position of Kerbin the mouse is 1.6m east and 1.77m north of his original starting position, which is 2.39 m from his starting position at an angle of 47.9 degrees north of east.

The Equilibrant Vector

The **equilibrant** of a force vector or set of force vectors is a single force vector which is exactly equal in magnitude and opposite in direction to the original vector or sum of vectors. The equilibrant, in effect, "cancels out" the original vector(s), or brings the set of vectors into equilibrium. To find an equilibrant, first find the resultant of the original vectors. The equilibrant is the opposite of the resultant you found!

2.24 Q: The diagram below represents two concurrent forces.

Which vector represents the force that will produce equilibrium with these two forces?

| (1) | (2) | (3) | (4) |

2.24 A: (3) The resultant of the two vectors would point up and to the left, therefore the equilibrant must point in the opposite direction, down and to the right.

Chapter 3: Kinematics

"I like physics. I think it is the best science out of all three of them, because generally it's more useful. You learn about speed and velocity and time, and that's all clever stuff."

— *Tom Felton*

Objectives

1. Calculate the kinetic energy of an object.
2. Describe the motion of an object qualitatively, mathematically, and graphically.
3. Design an experiment and analyze data to investigate the motion of an object.
4. Utilize center of mass of a two-object system to analyze the motion of a system.
5. Create and utilize both mathematical and graphical models to analyze the relationships between acceleration, velocity, and position of the center of mass of a system.
6. Sketch the theoretical path of a projectile.
7. Solve problems involving projectile motion for projectiles fired horizontally and at an angle.
8. Solve problems involving changing frames of reference and relative velocities.

Physics is all about energy in the universe, in all its various forms. Here on Earth, the source of all energy, directly or indirectly, is the conversion of mass into energy. We receive most of this energy from the sun. Solar power, wind power, hydroelectric power, fossil fuels, all can be traced back to the sun and the conversion of mass into energy. So where do you start in your study of the universe?

Theoretically, you could start by investigating any of these types of energy. In reality, however, by starting with energy of motion (also known as **kinetic energy**, K), you can develop a set of analytical problem solving skills from basic principles that will serve you well as you expand into the study of other types of energy.

For an object to have kinetic energy, it must be moving. Specifically, the kinetic energy of an object is equal to one half of the object's mass multiplied by the square of its velocity.

$$K = \tfrac{1}{2}mv^2$$

If kinetic energy is energy of motion, and energy is the ability or capacity to do work (moving an object), then you can think of kinetic energy as the ability or capacity of a moving object to move another object.

But what does it mean to be in motion? A moving object has a varying position. Its location changes as a function of time. So to understand kinetic energy, you'll need to better understand position and how position changes. This will lead into the first major unit, kinematics, from the Greek word kinein, meaning to move. Formally, kinematics is the branch of physics dealing with the description of an object's motion, leaving the study of the "why" of motion to the next major topic, dynamics.

Position, Distance, and Displacement

An object's **position** refers to its location at any given point in time. Position is a vector, and its magnitude is given by the symbol x. If we confine our study to motion in one dimension, we can define how far an object travels from its initial position as its **distance** (d). Distance, as defined by physics, is a **scalar**. It has a magnitude, or size, only. The basic unit of distance is the meter (m).

3.1 Q: On a sunny afternoon, a deer walks 1300 meters east to a creek for a drink. The deer then walks 500 meters west to the berry patch for dinner, before running 300 meters west when startled by a loud raccoon. What distance did the deer travel?

3.1 A: The deer traveled 1300m + 500m + 300m, for a total distance traveled (d) of 2100m.

Besides distance, in physics it is also helpful to know how far an object is from its starting point, or its change in position. The vector quantity **displacement** ($x - x_0$), or Δx, describes how far an object is from its starting point. The direction of the displacement vector points from the starting point to the finishing point.

Of special note is the symbolism for Δx. The delta symbol (Δ) indicates a change in a quantity, which is always the initial quantity subtracted from the final quantity.

Subscripts are used in physics to specify values of a quantity at a given point in time or space. The quantity x_0, then, refers to the position of an object at time t=0, typically referred to as the initial position. The final, or current, position is typically written without a subscript. Therefore, the displacement, written as Δx (which is always the final value subtracted from the initial value, or $x - x_0$), is the final position of the object minus the initial position of the object. Like distance, the units of displacement are meters.

3.2 Q: A deer walks 1300 m east to a creek for a drink. The deer then walked 500 m west to the berry patch for dinner, before running 300 m west when startled by a loud raccoon. What is the deer's displacement?

3.2 A: The deer's displacement was 500m east.

3.3 Q: Which are vector quantities?
(A) speed
(B) position
(C) mass
(D) displacement

3.3 A: (B and D) Position and displacement are vector quantities; they
have magnitude and direction.

3.4 Q: A student on her way to school walks four blocks east, three blocks
north, and another four blocks east, as shown in the diagram.

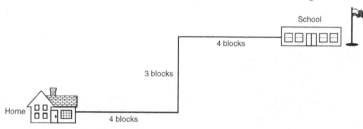

Compared to the distance she walks, the magnitude of her displacement from home to school is

(A) less

(B) greater

(C) the same

3.4 A: (A) The magnitude of displacement is always less than or equal
to the distance traveled.

3.5 Q: A hiker walks 5 kilometers due north and then 7 kilometers due east. What is the magnitude of her resultant
displacement? What total distance has she traveled?

3.5 A:

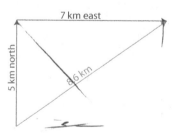

$$a^2 + b^2 = c^2 \rightarrow c = \sqrt{a^2 + b^2} = \sqrt{(5km)^2 + (7km)^2} = 8.6km$$

The hiker's resultant displacement is 8.6 km north of east.

The hiker's distance traveled is 12 kilometers.

Notice how for the exact same motion, distance and displacement have significantly different values based on their scalar or vector nature. Understanding the similarities (and differences) between these concepts is an important step toward understanding kinematics.

Speed and Velocity

Knowing only an object's distance and displacement doesn't tell the whole story. Going back to the deer example, there's a significant difference in the picture of the deer's afternoon if the deer's travels occurred over 5 minutes (300 seconds) as opposed to over 50 minutes (3000 seconds).

How exactly does the picture change? In order to answer that question, you'll need to understand some new concepts: average speed and average velocity.

Average speed, given the symbol $\bar{v}$, is defined as distance traveled divided by time, and it tells you the rate at which an object's distance traveled changes. When applying the formula, you must make sure that d is used to represent distance traveled.

$$\bar{v} = \frac{d}{t}$$

3.6 Q: A deer walks 1300 m east to a creek for a drink. The deer then walked 500 m west to the berry patch for dinner, before running 300 m west when startled by a loud raccoon. What is the deer's average speed if the entire trip took 600 seconds (10 minutes)?

3.6 A: $\bar{v} = \frac{d}{t} = \frac{2100m}{600s} = 3.5 \, ^{m}/_{s}$

Average velocity, also given the symbol $\bar{v}$, is defined as displacement, or change in position, over time. It tells you the rate at which an object's displacement, or position, changes. To calculate the vector quantity average velocity, you divide the vector quantity displacement by time. Note that if you want to find instantaneous speed or velocity, you must take the limit as the time interval gets extremely small (approaches 0).

$$\bar{v} = \frac{x - x_0}{t} = \frac{\Delta x}{t}$$

3.7 Q: A deer walks 1300 m east to a creek for a drink. The deer then walked 500 m west to the berry patch for dinner, before running 300 m west when startled by a loud raccoon. What is the deer's average velocity if the entire trip took 600 seconds (10 minutes)?

3.7 A: $\overline{v} = \dfrac{\Delta x}{t} = \dfrac{500m}{600s} = 0.83 \, m/_s$ east

Again, notice how you get very different answers for average speed compared to average velocity. The difference is realizing that distance and speed are scalars, and displacement and velocity are vectors. One way to help you remember these: **s**peed is a **s**calar, and **v**elocity is a **v**ector.

3.8 Q: Chuck the hungry squirrel travels 4m east and 3m north in search of an acorn. The entire trip takes him 20 seconds. Find: Chuck's distance traveled, Chuck's displacement, Chuck's average speed, and Chuck's average velocity.

3.8 A: $d = 4m + 3m = 7m$

$\Delta x = \sqrt{(4m)^2 + (3m)^2} = 5m \ northeast$

$\overline{v}(\text{avg. speed}) = \dfrac{d}{t} = \dfrac{7m}{20s} = 0.35 \, m/_s$

$\overline{v}(\text{avg. velocity}) = \dfrac{\Delta x}{t} = \dfrac{5m}{20s} = 0.25 \, m/_s \ northeast$

3.9 Q: On a highway, a car is driven 80 kilometers during the first 1.00 hour of travel, 50 kilometers during the next 0.50 hour, and 40 kilometers in the final 0.50 hour. What is the car's average speed for the entire trip?

(A) 45 km/h

(B) 60 km/h

(C) 85 km/h

(D) 170 km/h

3.9 A: (C) $\overline{v} = \dfrac{d}{t} = \dfrac{170km}{2h} = 85 \, km/_h$

3.10 Q: A person walks 150 meters due east and then walks 30 meters due west. The entire trip takes the person 10 minutes. Determine the magnitude and the direction of the person's total displacement.

3.10 A: 120m due east

3.11 Q: An athlete runs 3 kilometers at a constant speed of 5 meters per second and then 7 kilometers at a constant speed of 14.4 km/hr. What is the average speed of the runner during her 10 km run?

(A) 0.52 m/s

(B) 1.94 m/s

(C) 4.26 km/hr

(D) 15.3 km/hr

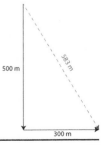

3.11 A: (D) Find the time for each leg of the run to determine the total time of the run.

$$t_1 = \frac{d}{v} = \frac{3000m}{5\,^m/_s} = 600s$$

$$t_2 = \frac{d}{v} = \frac{7km}{14.4\,^{km}/_{hr}} = 0.486hr = 1750s$$

$$\overline{v}_{total} = \frac{d_{total}}{t_{total}} = \frac{10,000m}{600s + 1750s} = 4.26\,^m/_s$$

Since 4.26 m/s is not one of the given answers, convert 4.26 m/s into km/hr to see if it matches an available answer.

$$4.26\,^m/_s \times \frac{1km}{1000m} \times \frac{3600s}{1hr} = 15.3\,^{km}/_{hr}$$

3.12 Q: A motorcycle travels 500 m south in 90 seconds, then 300 m east in 60 seconds. What is the magnitude of the motorcycle's average velocity?

(A) 1.33 m/s

(B) 3.89 m/s

(C) 5.33 m/s

(D) 6.48 m/s

3.12 A: (B) $\overline{v} = \frac{\Delta x}{t} = \frac{583m}{150s} = 3.89\,^m/_s$

To determine the average velocity, find the total displacement of the motorcycle as 583 m using the Pythagorean Theorem.

500 m

583 m

300 m

3.13 Q: A swimmer swims three-fifths the width of a river at one velocity (v), then swims the remainder of the river at half her initial velocity (½v). What was her average speed swimming across the river?

(A) 0.71v

(B) 0.75v

(C) 0.80v

(D) 0.88v

3.13 A: (A) Begin by finding the time for each leg of the journey, then determine the average velocity for the entire trip across the width of the river (r).

$$t_A = \frac{d}{v} = \frac{0.6r}{v}$$

$$t_B = \frac{d}{v} = \frac{0.4r}{0.5v} = \frac{0.8r}{v}$$

$$\bar{v} = \frac{d}{t} = \frac{r}{t_A + t_B} = \frac{r}{0.6\frac{r}{v} + 0.8\frac{r}{v}} = 0.71v$$

3.14 Q: Rush, the crime-fighting superhero, can run at a maximum speed of 30 m/s, while Evil Eddie, the criminal mastermind, can run 5 m/s. If Evil Eddie is 500 meters ahead of Rush:

A) How much time does Evil Eddie have to devise an escape plan?

B) How far must Rush run to capture Evil Eddie?

3.14 A: Start by setting up the problem. Let x_R represent Rush's distance from the origin, and x_{EE} represent Evil Eddie's distance from the origin. You can then model their positions as a function of time as:

$$x_R = 30t$$

$$x_{EE} = 500 + 5t$$

A) Evil Eddie runs out of time when they have the same position, so set x_R equal to x_{EE} in order to solve for the time when they meet.

$$x_R = x_{EE} \rightarrow 30t = 500 + 5t \rightarrow t = 20s$$

B) If Rush catches Evil Eddie in 20s, you can find the distance Rush must run by multiplying his maximum velocity by the time he must run.

$$x_R = 30t = (30 \tfrac{m}{s})(20s) = 600m$$

Acceleration

So you're starting to get a pretty good understanding of motion. But what would the world be like if velocity never changed? Objects at rest would remain at rest. Objects in motion would remain in motion at a constant speed and direction. Kinetic energy would never change (recall $K = \frac{1}{2}mv^2$?). It'd make for a pretty boring world. Thankfully, velocity can change, and this change in velocity leads to an **acceleration**.

More accurately, acceleration is the rate of change of velocity with respect to time. You can write this as:

$$a = \frac{\Delta v}{t} = \frac{v - v_0}{t}$$

This formula indicates that the change in velocity divided by the time interval gives you the average acceleration. Much like displacement and velocity, acceleration is a vector – it has a direction. Further, the units of acceleration are meters per second per second, or [m/s²]. Although it sounds complicated, all the units mean is that velocity changes at the rate of one meter per second, every second. So an object starting at rest and accelerating at 2 m/s² would be moving at 2 m/s after one second, 4 m/s after two seconds, 6 m/s after three seconds, and so on.

3.15 Q: Monty the Monkey accelerates uniformly from rest to a velocity of 9 m/s in a time span of 3 seconds. Calculate Monty's acceleration.

3.15 A: $a = \frac{\Delta v}{t} = \frac{v - v_0}{t} = \frac{9\,^m\!/_s - 0\,^m\!/_s}{3s} = 3\,^m\!/_{s^2}$

3.16 Q: Velocity is to speed as displacement is to
(A) acceleration
(B) time
(C) momentum
(D) distance

3.16 A: (D) distance. Velocity is the vector equivalent of speed, and displacement is the vector equivalent of distance.

The definition of acceleration can be rearranged to provide a relationship between velocity, acceleration, and time as follows:

$$a = \frac{\Delta v}{t} = \frac{v - v_0}{t}$$

$$v - v_0 = at$$

$$v = v_0 + at$$

This general formula for velocity can be utilized in any single direction. For linear motion in the x-direction, for example, the subscript x may be added to appropriate physical quantities to differentiate from motion in other directions (such as the x- and y- directions) as shown below.

$$v_x = v_{x0} + a_x t$$

The AP Physics 1 course will focus on motion in one dimension, though the basic principles of motion (and these equations) can be generalized to analyze and describe motion in multiple dimensions.

3.17 Q: The instant before a batter hits a 0.14-kilogram baseball, the velocity of the ball is 45 meters per second west. The instant after the batter hits the ball, the ball's velocity is 35 meters per second east. The bat and ball are in contact for 1.0×10^{-2} second. Determine the magnitude and direction of the average acceleration of the baseball while it is in contact with the bat.

3.17 A: Given: Find:

$v_0 = -45\,^m/_s$ a

$v = 35\,^m/_s$

$t = 1 \times 10^{-2}\,s$

$a = \dfrac{\Delta v}{t} = \dfrac{v - v_0}{t} = \dfrac{35\,^m/_s - (-45\,^m/_s)}{1 \times 10^{-2}\,s}$

$a = 8000\,^m/_{s^2}$ east

Because acceleration is a vector and has direction, it's important to realize that positive and negative values for acceleration indicate direction only. Take a look at some examples. First, an acceleration of zero implies an object moves at a constant velocity, so a car traveling at 30 m/s east with zero acceleration remains in motion at 30 m/s east.

If the car starts at rest and the car is given a positive acceleration of 5 m/s^2 east, the car speeds up as it moves east, going faster and faster each second. After one second, the car is traveling 5 m/s east. After two seconds, the car travels 10 m/s east. After three seconds, the car travels 15 m/s east, and so on.

But what happens if the car starts with a velocity of 15 m/s east, and it accelerates at a rate of -5 m/s^2 (or equivalently, 5 m/s^2 west)? The car will slow down as it moves to the east until its velocity becomes zero, then it will speed up as it continues to the west.

Positive accelerations don't necessarily indicate an object speeding up, and negative accelerations don't necessarily indicate an object slowing down. In one dimension, for example, if you call east the positive direction, a negative acceleration would indicate an acceleration vector pointing to the west. If the object is moving to the east (has a positive velocity), the negative acceleration would indicate the object is slowing down. If, however, the object is moving to the west (has a negative velocity), the negative acceleration would indicate the object is speeding up as it moves west.

Exasperating, isn't it? Putting it much more simply, if acceleration and velocity have the same sign (vectors in the same direction), the object is speeding up. If acceleration and velocity have opposite signs (vectors in opposite directions), the object is slowing down.

Particle Diagrams

Graphs and diagrams are terrific tools for understanding physics, and they are especially helpful for studying motion, a phenomenon that we are used to perceiving visually. Particle diagrams, sometimes referred to as ticker-tape diagrams or dot diagrams, show the position or displacement of an object at evenly spaced time intervals.

Think of a particle diagram like an oil drip pattern... if a car has a steady oil drip, where one drop of oil falls to the ground every second, the pattern of the oil droplets on the ground could represent the motion of the car with respect to time. By examining the oil drop pattern, a bystander could draw conclusions about the displacement, velocity, and acceleration of the car, even if he wasn't able to watch the car drive by! The oil drop pattern is known as a particle, or ticker-tape, diagram.

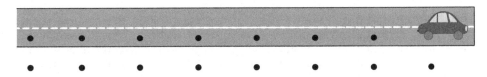

From the particle diagram above you can see that the car was moving either to the right or the left, and since the drops are evenly spaced, you can say with certainty that the car was moving at a constant velocity, and since velocity isn't changing, acceleration must be 0. So what would the particle diagram look like if the car was accelerating to the right? Take a look below and see!

The oil drops start close together on the left, and get further and further apart as the object moves toward the right. Of course, this pattern could also have been produced by a car moving from right to left, beginning with a high velocity at the right and slowing down as it moves toward the left. Because the velocity vector (pointing to the left) and the acceleration vector (pointing to the right) are in opposite directions, the object slows down. This is a case where, if you called to the right the positive direction, the car would have a negative velocity, a positive acceleration, and it would be slowing down. Check out the resulting particle diagram below!

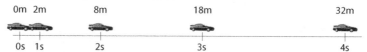

Can you think of a case in which the car could have a negative velocity and a negative acceleration, yet be speeding up? Draw a picture of the situation!

3.18 Q: The diagram below shows a car beginning from rest and accelerating uniformly down the road.

| 0m 2m | 8m | 18m | 32m |

| 0s 1s | 2s | 3s | 4s |

What is the average speed of the car between 2 and 4 seconds?

3.18 A: $\bar{v} = \dfrac{d}{t} = \dfrac{24m}{2s} = 12 \, m/s$

Position-Time (x-t) Graphs

As you've observed, particle diagrams can help you understand an object's motion, but they don't always tell you the whole story. You'll have to investigate some other types of motion graphs to get a clearer picture.

The position-time graph shows the displacement (or, in the case of scalar quantities, distance) of an object as a function of time. Positive displacements indicate the object's position is in the positive direction from its starting point, while negative displacements indicate the object's position is opposite the positive direction. Let's look at a few examples.

Chapter 3: Kinematics

Suppose Stella the Wonder Dog wanders away from her house at a constant velocity of 1 m/s, stopping only when she's five meters away (which, of course, takes five seconds). She then decides to take a short five second rest in the grass. After her five second rest, she hears the dinner bell ring, so she runs back to the house at a speed of 2 m/s. The position-time graph for her motion would look something like this:

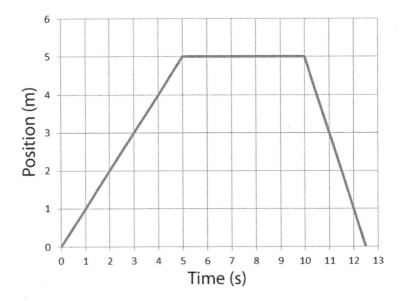

As you can see from the plot, Stella's displacement begins at zero meters at time zero. Then, as time progresses, Stella's position changes at a rate of 1 m/s, so that after one second, Stella is one meter away from her starting point. After two seconds, she's two meters away, and so forth, until she reaches her maximum displacement of five meters from her starting point at a time of five seconds (her position is now $x=5$ m). Stella then remains at that position for 5 seconds while she takes a rest. Following her rest, at time $t=10$ seconds, Stella hears the dinner bell and races back to the house at a speed of 2 m/s, so the graph ends when Stella returns to her starting point at the house, a total distance traveled of 10m, and a total displacement of zero meters.

As you look at the position-time graph, notice that at the beginning, when Stella is moving in a positive direction, the graph has a positive slope. When the graph is flat (has a zero slope), Stella is not moving. When the graph has a negative slope, Stella is moving in the negative direction. It's also easy to see that the steeper the slope of the graph, the faster Stella is moving.

3.19 Q: The graph below represents the displacement of an object moving in a straight line as a function of time.

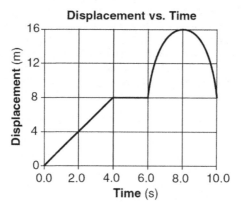

Displacement vs. Time

What was the total distance traveled by the object during the 10-second time interval?

3.19 A: Total distance traveled is 8 meters forward from 0 to 4 seconds, then 8 meters forward from 6 to 8 seconds, then 8 meters backward from 8 to 10 seconds, for a total of 24 meters.

Velocity-Time (v-t) Graphs

Just as important to understanding motion is the velocity-time graph, which shows the velocity of an object on the y-axis, and time on the x-axis. Positive values indicate velocities in the positive direction, while negative values indicate velocities in the opposite direction. In reading these graphs, it's important to realize that a straight horizontal line indicates the object maintaining a constant velocity – it can still be moving, its velocity just isn't changing. A value of 0 on the v-t graph indicates the object has come to a stop. If the graph crosses the x-axis, the object was moving in one direction, came to a stop, and switched the direction of its motion. Let's look at the v-t graph for Stella the Wonder Dog's Adventure:

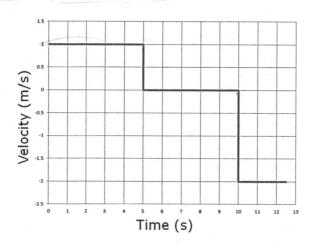

For the first five seconds of Stella's journey, you can see she maintains a constant velocity of 1 m/s. Then, when she stops to rest, her velocity changes to zero for the duration of her rest. Finally, when she races back to the house for dinner, she maintains a negative velocity of 2 m/s. Because velocity is a vector, the negative sign indicates that Stella's velocity is in the opposite direction (initially the direction away from the house was positive, so back toward the house must be negative!).

As I'm sure you can imagine, the position-time graph of an object's motion and the velocity-time graph of an object's motion are closely related. You'll explore these relationships next.

Graph Transformations

In looking at a position-time graph, the faster an object's position/displacement changes, the steeper the slope of the line. Since velocity is the rate at which an object's position changes, the slope of the position-time graph at any given point in time gives you the velocity at that point in time. You can obtain the slope of the position-time graph using the following formula:

$$slope = \frac{rise}{run} = \frac{y_2 - y_1}{x_2 - x_1}$$

Realizing that the rise in the graph is actually Δx, and the run is Δt, you can substitute these variables into the slope equation to find:

$$slope = \frac{rise}{run} = \frac{\Delta x}{\Delta t} = v$$

With a little bit of interpretation, it's easy to show that the slope is really just change in position over time, which is the definition of velocity. Put directly, the slope of the position-time graph gives you the velocity.

Of course, it only makes sense that if you can determine velocity from the position-time graph, you should be able to work backward to determine change in position (displacement) from the v-t graph. If you have a v-t graph, and you want to know how much an object's position changed in a time interval, take the area under the curve within that time interval.

So, if taking the slope of the position-time graph gives you the rate of change of position, which is called velocity, what do you get when you take the slope of the v-t graph? You get the rate of change of velocity, which is called acceleration! The slope of the v-t graph, therefore, tells you the acceleration of an object.

$$slope = \frac{rise}{run} = \frac{\Delta v}{\Delta t} = a$$

3.20 Q: The graph below represents the motion of a car during a 6.0-second time interval.

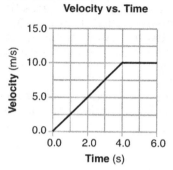

Velocity vs. Time

(A) What is the total distance traveled by the car during this 6-second interval?

(B) What is the acceleration of the car at t = 5 seconds?

3.20 A: (A) distance = area under graph

$$distance = Area_{triangle} + Area_{rectangle}$$

$$distance = \frac{1}{2}bh + lw$$

$$distance = \frac{1}{2}(4s)(10\,{}^{m}\!/_{s}) + (2s)(10\,{}^{m}\!/_{s})$$

$$distance = 40m$$

(B) acceleration = slope at t=5 seconds = 0 because graph is flat at t=5 seconds.

3.21 Q: The graph below represents the velocity of an object traveling in a straight line as a function of time.

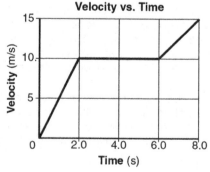

Velocity vs. Time

Determine the magnitude of the total displacement of the object at the end of the first 6.0 seconds.

3.21 A: displacement = area under graph

displacement = $Area_{triangle} + Area_{rectangle}$

displacement = $\frac{1}{2}bh + lw$

displacement = $\frac{1}{2}(2s)(10\,{}^{m}\!/_{s}) + (4s)(10\,{}^{m}\!/_{s})$

displacement = $50m$

3.22 Q: The graph below shows the velocity of a race car moving along a straight line as a function of time.

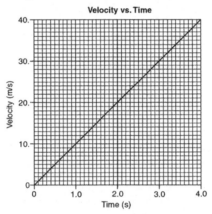

What is the magnitude of the displacement of the car from t = 2.0 seconds to t = 4.0 seconds?

(A) 20 m

(B) 40 m

(C) 60 m

(D) 80 m

3.22 A: (C) 60 m

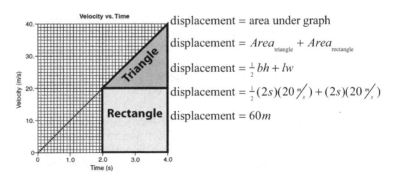

displacement = area under graph

displacement = $Area_{triangle} + Area_{rectangle}$

displacement = $\frac{1}{2}bh + lw$

displacement = $\frac{1}{2}(2s)(20\,{}^{m}\!/_{s}) + (2s)(20\,{}^{m}\!/_{s})$

displacement = $60m$

3.23 Q: The displacement-time graph below represents the motion of a cart initially moving forward along a straight line.

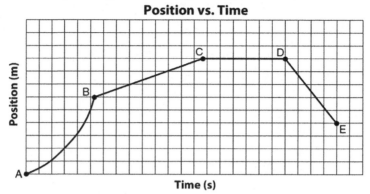

Position vs. Time

During which interval is the cart moving forward at constant speed?

(A) AB

(B) BC

(C) CD

(D) DE

3.23 A: (B) The slope of the position-time graph is constant and positive during interval BC, therefore the velocity of the cart must be constant and positive in that interval. Although the cart is moving forward during interval AB, the slope of the position-time graph is increasing, indicating that its speed is increasing, making this an incorrect choice. During DE, the cart is move backward at constant speed, making this another incorrect choice.

Acceleration-Time (a-t) Graphs

Much like velocity, you can make a graph of acceleration vs. time by plotting the rate of change of an object's velocity (its acceleration) on the y-axis, and placing time on the x-axis.

When you took the slope of the position-time graph, you obtained the object's velocity. In the same way, taking the slope of the v-t graph gives you the object's acceleration. Going the other direction, when you analyzed the v-t graph, you found that taking the area under the v-t graph provided you with information about the object's change in position. In similar fashion, taking the area under the a-t graph tells you how much an object's velocity changes.

Putting it all together, you can go from position-time to velocity-time by taking the slope, and you can go from velocity-time to acceleration-time by taking the slope. Or, going the other direction, the area under the acceleration-time curve gives you an object's change in velocity, and the area under the velocity-time curve gives you an object's change in position.

Graphical Analysis of Motion
How do I move from one type of graph to another?

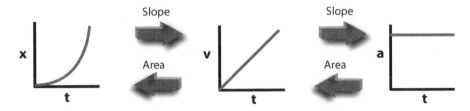

3.24 Q: Which graph best represents the motion of a block accelerating uniformly down an inclined plane?

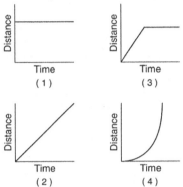

3.24 A: (4) If the block accelerates uniformly, it must have a constant acceleration. This means the v-t graph must be a straight line, since its slope, which is equal to its acceleration, must be constant. Therefore, the v-t graph must look something like the graph at right. The slope of the position-time graph, which gives velocity, must be constantly increasing to give the v-t graph at right. The only answer choice with a constantly increasing slope is (4), so (4) must be the answer!

3.25 Q: A student throws a baseball vertically upward and then catches it. If vertically upward is considered to be the positive direction, which graph best represents the relationship between velocity and time for the baseball?

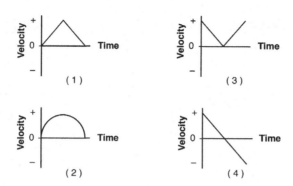

3.25 A: (4) If up is considered the positive direction, and the baseball is thrown upward to start its motion, it begins with a positive velocity. At its highest point, its vertical velocity becomes 0. Then, it speeds up as it comes down, so it obtains a larger and larger negative velocity.

3.26 Q: A cart travels with a constant nonzero acceleration along a straight line. Which graph best represents the relationship between the distance the cart travels and time of travel?

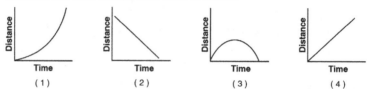

3.26 A: (1) A constant acceleration is caused by a linearly increasing velocity. Since velocity is obtained from the slope of the position-time graph, the position-time graph must be continually increasing to provide the correct v-t graph.

3.27 Q: Which of the following pairs of graphs best shows the distance traveled versus time and speed versus time for a car accelerating down a hill from rest?

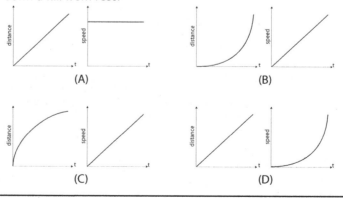

3.27 A: (B) speed is increasing at a constant rate while the rate at which distance increases constantly increases.

Kinematic Equations

Motion graphs such as the position-time, velocity-time, and acceleration-time graphs are terrific tools for understanding motion. However, there are times when graphing motion may not be the most efficient or effective way of understanding the motion of an object. To assist in these situations, you can add a set of problem-solving equations to your physics toolbox, known as the **kinematic equations**. These equations can help you solve for key variables describing the motion of an object when you have a *constant acceleration*. Once you know the value of any three variables, you can use the kinematic equations to solve for the other two!

Variable	Meaning
v_0	initial velocity
v	final velocity
Δx	displacement
a	acceleration
t	time

$$v_x = v_{x0} + a_x t$$

$$v_x^2 = v_{x0}^2 + 2a_x(x - x_0)$$

$$x = x_0 + v_{x0}t + \tfrac{1}{2}a_x t^2$$

In using these equations to solve motion problems, it's important to take care in setting up your analysis before diving into a solution. Key steps to solving kinematics problems include:

1. Labeling your analysis for horizontal (x-axis) or vertical (y-axis) motion.
2. Choosing and indicating a positive direction (typically the direction of initial motion).
3. Creating a motion analysis table (v_0, v, Δx, a, t). Note that Δx is a change in position, or displacement, and can be re-written as x-x_0.
4. Using what you know about the problem to fill in your "givens" in the table.
5. Once you know three items in the table, use kinematic equations to solve for any unknowns.
6. Verify that your solution makes sense.

Take a look at a sample problem to see how this strategy can be employed.

3.28 Q: A race car starting from rest accelerates uniformly at a rate of 4.90 meters per second². What is the car's speed after it has traveled 200 meters?

(A) 1960 m/s

(B) 62.6 m/s

(C) 44.3 m/s

(D) 31.3 m/s

3.28 A: Step 1: Horizontal Problem

Step 2: Positive direction is direction car starts moving.

Step 3 & 4:

Variable	Value
v_0	0 m/s
v	FIND
Δx	200 m
a_x	4.90 m/s²
t	?

Step 5: Choose a kinematic equation that includes the given information and the information sought, and solve for the unknown showing the initial formula, substitution with units, and answer with units.

$$v^2 = v_0^2 + 2a\Delta x$$
$$v^2 = (0\,{}^m\!/_s)^2 + 2(4.90\,{}^m\!/_{s^2})(200m)$$
$$v^2 = 1960\,{}^{m^2}\!/_{s^2}$$
$$v^2 = \sqrt{1960\,{}^{m^2}\!/_{s^2}} = 44.3\,{}^m\!/_s$$

Step 6: (C) 44.3 m/s is one of the given answer choices, and is a reasonable speed for a race car (44.3 m/s is approximately 99 miles per hour).

This strategy also works for vertical motion problems.

3.29 Q: An astronaut standing on a platform on the Moon drops a hammer. If the hammer falls 6.0 meters vertically in 2.7 seconds, what is its acceleration?

(A) 1.6 m/s²

(B) 2.2 m/s²

(C) 4.4 m/s²

(D) 9.8 m/s²

3.29 A: Step 1: Vertical Problem

Step 2: Positive direction is down (direction hammer starts moving.)

Step 3 & 4: Note that a dropped object has an initial vertical velocity of 0 m/s.

Variable	Value
v_0	0 m/s
v	?
Δy	6 m
a	FIND
t	2.7 s

Step 5: Choose a kinematic equation that includes the given information and the information sought, and solve for the unknown showing the initial formula, substitution with units, and answer with units.

$$\Delta y = v_0 t + \tfrac{1}{2} at^2 \xrightarrow{v_0 = 0}$$

$$\Delta y = \tfrac{1}{2} at^2$$

$$a = \frac{2\Delta y}{t^2}$$

$$a = \frac{2(6m)}{(2.7s)^2} = 1.6 \, m/_{s^2}$$

Step 6: (A) 1.6 m/s² is one of the given answer choices, and is less than the acceleration due to gravity on the surface of the Earth (9.8 m/s²). This answer can be verified further by searching on the Internet to confirm that the acceleration due to gravity on the surface of the moon is indeed 1.6 m/s².

In some cases, you may not be able to solve directly for the "find" quantity. In these cases, you can solve for the other unknown variable first, then choose an equation to give you your final answer.

3.30 Q: A car traveling on a straight road at 15.0 meters per second accelerates uniformly to a speed of 21.0 meters per second in 12.0 seconds. The total distance traveled by the car in this 12.0-second time interval is

(A) 36.0 m

(B) 180 m

(C) 216 m

(D) 252 m

3.30 A: Horizontal Problem, positive direction is forward.

Variable	Value
v_0	15 m/s
v	21 m/s
Δx	FIND
a	?
t	12 s

Can't find Δx directly, find a first.

$$v = v_0 + at$$

$$a = \frac{v - v_0}{t}$$

$$a = \frac{21\,\text{m}/\text{s} - 15\,\text{m}/\text{s}}{12s} = 0.5\,\text{m}/\text{s}^2$$

Now solve for Δx.

$$\Delta x = v_0 t + \tfrac{1}{2} a t^2$$

$$\Delta x = (15\,\text{m}/\text{s})(12s) + \tfrac{1}{2}(0.5\,\text{m}/\text{s}^2)(12s)^2$$

$$\Delta x = 216m$$

Check: (C) 216m is a given answer, and is reasonable, as this is greater than the 180m the car would have traveled if remaining at a constant speed of 15 m/s for the 12 second time interval.

3.31 Q: A car initially traveling at a speed of 16 meters per second accelerates uniformly to a speed of 20 meters per second over a distance of 36 meters. What is the magnitude of the car's acceleration?

(A) 0.11 m/s²

(B) 2.0 m/s²

(C) 0.22 m/s²

(D) 9.0 m/s²

3.31 A: (B) $v^2 = v_0^2 + 2a\Delta x$

$$a = \frac{v^2 - v_0^2}{2\Delta x}$$

$$a = \frac{(20\,\text{m}/\text{s})^2 - (16\,\text{m}/\text{s})^2}{2(36m)} = 2\,\text{m}/\text{s}^2$$

3.32 Q: An astronaut drops a hammer from 2.0 meters above the surface of the Moon. If the acceleration due to gravity on the Moon is 1.62 meters per second², how long will it take for the hammer to fall to the Moon's surface?

(A) 0.62 s

(B) 1.2 s

(C) 1.6 s

(D) 2.5 s

3.32 A: (C) $\Delta y = v_0 t + \frac{1}{2} a t^2 \xrightarrow{\;v_0=0\;}$

$$\Delta y = \frac{1}{2} a t^2$$

$$t = \sqrt{\frac{2\Delta y}{a}} = \sqrt{\frac{2(2m)}{1.62 \, {}^m\!/_{s^2}}} = 1.57s$$

3.33 Q: A car increases its speed from 9.6 meters per second to 11.2 meters per second in 4.0 seconds. The average acceleration of the car during this 4.0-second interval is

(A) 0.40 m/s²

(B) 2.4 m/s²

(C) 2.8 m/s²

(D) 5.2 m/s²

3.33 A: (A) $v = v_0 + at$

$$a = \frac{v - v_0}{t} = \frac{(11.2 \, {}^m\!/_s) - (9.6 \, {}^m\!/_s)}{4s} = 0.4 \, {}^m\!/_{s^2}$$

3.34 Q: The diagram below shows a car beginning from rest and accelerating uniformly down the road.

0m 2m   8m   18m   32m

0s   1s   2s   3s   4s

A) What is the car's acceleration during the first three seconds of its journey?

B) What is its instantaneous velocity when it has traveled a distance of 24 meters?

3.34 A: A) $x = x_0 + v_{x0} t + \frac{1}{2} a_x t^2 \xrightarrow[\;v_{x0}=0\;]{\;x_0=0\;} x = \frac{1}{2} a_x t^2 \rightarrow$

$$a_x = \frac{2x}{t^2} = \frac{2(18m)}{(3s)^2} = 4 \, {}^m\!/_{s^2}$$

B) $v_x^2 = v_{x0}^2 + 2 a_x (x - x_0) = 2(4 \, {}^m\!/_{s^2})(24m - 0) \rightarrow$

$$v_x^2 = 192 \, {}^{m^2}\!/_{s^2} \rightarrow v_x = 13.9 \, {}^m\!/_s$$

3.35 Q: A rock falls from rest a vertical distance of 0.72 meters to the surface of a planet in 0.63 seconds. The magnitude of the acceleration due to gravity on the planet is

(A) 1.1 m/s²

(B) 2.3 m/s²

(C) 3.6m/s²

(D) 9.8 m/s²

3.35 A: (C) $\Delta y = v_0 t + \frac{1}{2}at^2 \xrightarrow{v_0=0}$

$$\Delta y = \frac{1}{2}at^2$$

$$a = \frac{2\Delta y}{t^2} = \frac{2(0.72m)}{(0.63s)^2} = 3.6\,m/_{s^2}$$

3.36 Q: The speed of an object undergoing constant acceleration increases from 8.0 meters per second to 16.0 meters per second in 10 seconds. How far does the object travel during the 10 seconds?

(A) 3.6×10² m

(B) 1.6×10² m

(C) 1.2×10² m

(D) 8.0×10¹ m

3.36 A: (C) Can't find Δx directly, find a first.

$$v = v_0 + at$$

$$a = \frac{v - v_0}{t}$$

$$a = \frac{16\,m/_s - 8\,m/_s}{10s} = 0.8\,m/_{s^2}$$

Now solve for Δx.

$$\Delta x = v_0 t + \frac{1}{2}at^2$$

$$\Delta x = (8\,m/_s)(10s) + \frac{1}{2}(0.8\,m/_{s^2})(10s)^2$$

$$\Delta x = 120m$$

Free Fall

Examination of free-falling bodies dates back to the days of Aristotle. At that time Aristotle believed that more massive objects would fall faster than less massive objects. He believed this in large part due to the fact that when examining a rock and a feather falling from the same height it is clear that the rock hits the ground first. Upon further examination it is clear that Aristotle was incorrect in his hypothesis.

Chapter 3: Kinematics

As proof, take a basketball and a piece of paper. Drop them simultaneously from the same height... do they land at the same time? Probably not. Now take that piece of paper and crumple it up into a tight ball and repeat the experiment. Now what do you see happen? You should see that both the ball and the paper land at the same time. Therefore you can conclude that Aristotle's predictions did not account for the effect of air resistance. For the purposes of this course, drag forces such as air resistance will be neglected.

In the 17th century, Galileo Galilei began a re-examination of the motion of falling bodies. Galileo, recognizing that air resistance affects the motion of a falling body, executed his famous thought experiment in which he continuously asked what would happen if the effect of air resistance was removed. Commander David Scott of Apollo 15 performed this experiment while on the moon. He simultaneously dropped a hammer and a feather, and observed that they reached the ground at the same time.

Since Galileo's experiments, scientists have come to a better understanding of how the gravitational pull of the Earth accelerates free-falling bodies. Through experimentation it has been determined that the local **gravitational field strength (g)** on the surface of the Earth is 9.8 N/kg, which further indicates that all objects in free fall (neglecting air resistance) experience an equivalent acceleration of 9.8 m/s^2 toward the center of the Earth. (NOTE: If you move off the surface of the Earth the local gravitational field strength, and therefore the acceleration due to gravity, changes.) For the purposes of the AP exam, it is acceptable to estimate g as 10 m/s^2 to simplify calculations.

You can look at free-falling bodies as objects being dropped from some height or thrown vertically upward. In this examination you will analyze the motion of each condition.

Objects Falling From Rest

Objects starting from rest have an initial velocity of zero, giving you your first kinematic quantity needed for problem solving. Beyond that, if you call the direction of initial motion (down) positive, the object will have a positive acceleration and speed up as it falls.

An important first step in analyzing objects in free fall is deciding which direction along the y-axis you are going to call positive and which direction will therefore be negative. Although you can set your positive direction any way you want and get the correct answer, following the hints below can simplify your work to reach the correct answer consistently.

1. Identify the direction of the object's initial motion and assign that as the positive direction. In the case of a dropped object, the positive y-direction

will point toward the bottom of the paper.
2. With the axis identified you can now identify and write down your given kinematic information. Don't forget that a dropped object has an initial velocity of zero.
3. Notice the direction the vector arrows are drawn — if the velocity and acceleration point in the same direction, the object speeds up. If they point in opposite directions, the object slows down.

3.37 Q: What is the speed of a 2.5-kilogram mass after it has fallen freely from rest through a distance of 12 meters?

(A) 4.8 m/s
(B) 15 m/s
(C) 30 m/s
(D) 43 m/s

3.37 A: Vertical Problem: Declare down as the positive direction. This means that the acceleration, which is also down, is a positive quantity.

Variable	Value
v_0	0 m/s
v	FIND
Δy	12 m
a	9.8 m/s²
t	?

$$v^2 = v_0^2 + 2a\Delta y$$
$$v^2 = (0\,^m\!/_s)^2 + 2(9.8\,^m\!/_{s^2})(12m)$$
$$v^2 = 235\,^{m^2}\!/_{s^2}$$
$$v = \sqrt{235\,^{m^2}\!/_{s^2}} = 15.3\,^m\!/_s$$

Correct answer is (B) 15 m/s.

3.38 Q: How far will a brick starting from rest fall freely in 3.0 seconds?

(A) 15 m
(B) 29 m
(C) 44 m
(D) 88 m

3.38 A: (C) 44m

Variable	Value
v_0	0 m/s
v	?
Δy	FIND
a	9.8 m/s²
t	3 s

$$\Delta y = v_0 t + \tfrac{1}{2}at^2$$

$$\Delta y = (0\,m/s)(3s) + \tfrac{1}{2}(9.8\,m/s^2)(3s)^2$$

$$\Delta y = 44m$$

3.39 Q: A ball dropped from rest falls freely until it hits the ground with a speed of 20 meters per second. The time during which the ball is in free fall is approximately

(A) 1 s

(B) 2 s

(C) 0.5 s

(D) 10 s

3.39 A:

Variable	Value
v_0	0 m/s
v	20 m/s
Δy	?
a	9.8 m/s²
t	FIND

$$v = v_0 + at$$

$$t = \frac{v - v_0}{a}$$

$$t = \frac{20\,m/s - 0\,m/s}{9.8\,m/s^2} = 2.04s$$

(B) 2 s

Objects Launched Upward

Examining the motion of an object launched vertically upward is done in much the same way you examined the motion of an object falling from rest. The major difference is that you have to look at two segments of its motion instead of one: both up *and* down.

Before you get into establishing a frame of reference and work-ing through the quantitative analysis, you must build a solid conceptual understanding of what is happening while the ball is in the air. Consider the ball being thrown vertically into the air as shown in the diagram.

In order for the ball to move upwards, its initial velocity must be greater than zero. As the ball rises, its velocity decreases until it reaches its maximum height, where it stops and then begins to fall. As the ball falls, its speed increases. In other words, the ball is accelerating the entire time it is in the air: both on the way up, at the instant it stops at its highest point, and on the way down.

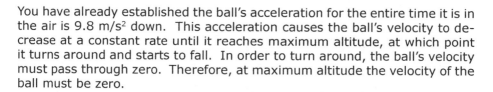

The cause of the ball's acceleration is gravity. The entire time the ball is in the air, its acceleration is 9.8 m/s² down provided this occurs on the surface of the Earth. Note that the accel-eration can be either 9.8 m/s² or -9.8 m/s². The sign of the acceleration depends on the direction you declared as positive, but in all cases the direction of the acceleration due to gravity is down, to-ward the center of the Earth.

You have already established the ball's acceleration for the entire time it is in the air is 9.8 m/s² down. This acceleration causes the ball's velocity to de-crease at a constant rate until it reaches maximum altitude, at which point it turns around and starts to fall. In order to turn around, the ball's velocity must pass through zero. Therefore, at maximum altitude the velocity of the ball must be zero.

3.40 Q: A ball thrown vertically upward reaches a maximum height of 30 meters above the surface of Earth. At its maximum height, the speed of the ball is

(A) 0 m/s

(B) 3.1 m/s

(C) 9.8 m/s

(D) 24 m/s

3.40 A: (A) 0 m/s. The instantaneous speed of any projectile launched vertically at its maximum height is zero.

Because gravity provides the same acceleration to the ball on the way up (slowing it down) as on the way down (speeding it up), the time to reach maximum altitude is the same as the time to return to its launch position. In similar fashion, the initial velocity of the ball on the way up will equal the velocity of the ball at the instant it reaches the point from which it was launched on the way down. Put another way, the time to go up is equal to the time to go down, and the initial velocity up is equal to the final velocity down (assuming the object begins and ends at the same height above ground).

Now that a conceptual understanding of the ball's motion has been established, you can work toward a quantitative solution. Following the rule of thumb established previously, you can start by assigning the direction the ball begins to move as positive. Remember that assigning positive and negative directions are completely arbitrary. You have the freedom to assign them how you see fit. Once you assign them, however, don't change them.

Once this positive reference direction has been established, all other velocities and displacements are assigned accordingly. For example, if up is the positive direction, the acceleration due to gravity will be negative, because the acceleration due to gravity points down, toward the center of the Earth. At its highest point, the ball will have a positive displacement, and will have a zero displacement when it returns to its starting point. If the ball isn't caught, but continues toward the Earth past its starting point, it will have a negative displacement.

A "trick of the trade" to solving free fall problems involves symmetry. The time an object takes to reach its highest point is equal to the time it takes to return to the same vertical position. The speed with which the projectile begins its journey upward is equal to the speed of the projectile when it returns to the same height (although, of course, its velocity is in the opposite direction). If you want to simplify the problem, vertically, at its highest point, the vertical velocity is 0. This added information can assist you in filling out your vertical motion table. If you cut the object's motion in half, you can simplify your problem solving – but don't forget that if you want the total time in the air, you must double the time it takes for the object to rise to its highest point.

3.41 Q: A basketball player jumped straight up to grab a rebound. If she was in the air for 0.80 seconds, how high did she jump?

(A) 0.50 m

(B) 0.78 m

(C) 1.2 m

(D) 3.1 m

3.41 A: Define up as the positive y-direction. Note that if basketball player is in the air for 0.80 seconds, she reaches her maximum height at a time of 0.40 seconds, at which point her velocity is zero.

Variable	Value
v_0	?
v	0 m/s
Δy	FIND
a	-9.8 m/s²
t	0.40 s

Can't solve for Δy directly with given information, so find v_0 first.

$$v = v_0 + at$$
$$v_0 = v - at$$
$$v_0 = 0 - (-9.8\,{}^{m}\!/_{s^2})(0.40s) = 3.92\,{}^{m}\!/_{s}$$

Now with v_0 known, solve for Δy.

$$\Delta y = v_0 t + \tfrac{1}{2}at^2$$
$$\Delta y = (3.92\,{}^{m}\!/_{s})(0.40s) + \tfrac{1}{2}(-9.8\,{}^{m}\!/_{s^2})(0.40s)^2$$
$$\Delta y = 0.78m$$

Correct answer is **(B)** 0.78 m. This is a reasonable height for a basketball player to jump.

If, instead, you analyze the motion on the way down, you find that v_0=0, a=9.8 m/s², and t=0.4 s. Solving for displacement:

$$\Delta y = v_0 t + \tfrac{1}{2}at^2 = (0)(0.4s) + \tfrac{1}{2}(9.8\,{}^{m}\!/_{s^2})(0.4s)^2 = 0.78m$$

You find the same answer, but with significantly less work!

3.42 Q: Which graph best represents the relationship between the acceleration of an object falling freely near the surface of Earth and the time that it falls?

(1)   (2)   (3)   (4)

3.42 A: (4) The acceleration due to gravity is a constant 9.8 m/s² down on the surface of the Earth.

3.43 Q: A ball is thrown straight downward with a speed of 0.50 meter per second from a height of 4.0 meters. What is the speed of the ball 0.70 seconds after it is released?

(A) 0.50 m/s

(B) 7.4 m/s

(C) 9.8 m/s

(D) 15 m/s

3.43 A: (B) 7.4 m/s. Note that in filling out the kinematics table, the height of 4 meters is not the displacement of the ball, but is extra unneeded information.

Variable	Value
v_0	0.50 m/s
v	FIND
Δy	?
a	9.8 m/s²
t	0.70 s

$$v = v_0 + at$$
$$v = 0.50 \, \tfrac{m}{s} + (9.8 \, \tfrac{m}{s^2})(0.70s)$$
$$v = 7.4 \, \tfrac{m}{s}$$

3.44 Q: A baseball dropped from the roof of a tall building takes 3.1 seconds to hit the ground. How tall is the building? [Neglect friction.]

(A) 15 m

(B) 30 m

(C) 47 m

(D) 94 m

3.44 A: (C) 47 m

$$\Delta y = v_0 t + \tfrac{1}{2}at^2$$
$$\Delta y = (0\tfrac{m}{s})(3.1s) + \tfrac{1}{2}(9.8\tfrac{m}{s^2})(3.1s)^2$$
$$\Delta y = 47m$$

3.45 Q: A 0.25-kilogram baseball is thrown upward with a speed of 30 meters per second. Neglecting friction, the maximum height reached by the baseball is approximately

(A) 15 m

(B) 46 m

(C) 74 m

(D) 92 m

3.45 A: (B) 46m

$$v^2 = v_0^2 + 2a\Delta y$$
$$\Delta y = \frac{v^2 - v_0^2}{2a}$$
$$\Delta y = \frac{(0\tfrac{m}{s})^2 - (30\tfrac{m}{s})^2}{2(-9.8\tfrac{m}{s^2})} = 45.9m$$

3.46 Q: Three model rockets of varying masses are launched vertically upward from the ground with varying initial velocities. From highest to lowest, rank the maximum height reached by each rocket. Neglect air resistance.

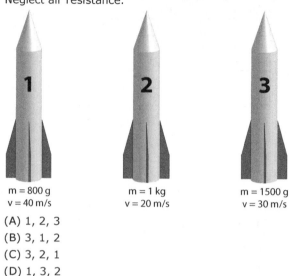

m = 800 g   m = 1 kg   m = 1500 g
v = 40 m/s   v = 20 m/s   v = 30 m/s

(A) 1, 2, 3

(B) 3, 1, 2

(C) 3, 2, 1

(D) 1, 3, 2

3.46 A: (D) 1, 3, 2. The acceleration due to gravity is independent of mass, therefore the object with the largest initial vertical velocity will reach the highest maximum height.

Projectile Motion

Projectile motion problems, or problems of an object launched in both the x- and y- directions, can be analyzed using the physics you already know. The key to solving these types of problems is realizing that the horizontal component of the object's motion is independent of the vertical component of the object's motion. Since you already know how to solve horizontal and vertical kinematics problems, all you have to do is put the two results together!

Start these problems by making separate motion tables for vertical and horizontal motion. Vertically, the setup is the same for projectile motion as it is for an object in free fall. Horizontally, gravity only pulls an object down, it never pulls or pushes an object horizontally; therefore the horizontal acceleration of any projectile is zero. If the acceleration horizontally is zero, velocity must be constant; therefore v_0 horizontally must equal v horizontally. Finally, to tie the problem together, realize that the time the projectile is in the air vertically must be equal to the time the projectile is in the air horizontally.

When an object is launched or thrown completely horizontally, such as a rock thrown horizontally off a cliff, the initial velocity of the object is its initial horizontal velocity. Because horizontal velocity doesn't change, this velocity is also the object's final horizontal velocity, as well as its average horizontal velocity. Further, the initial vertical velocity of the projectile is zero. This means that you could hurl an object 1000 m/s horizontally off a cliff, and simultaneously drop an object off the cliff from the same height, and they will both reach the ground at the same time (even though the hurled object has traveled a greater distance).

3.47 Q: Fred throws a baseball 42 m/s horizontally from a height of two meters. How far will the ball travel before it reaches the ground?

3.47 A: To solve this problem, you must first find how long the ball will remain in the air. This is a vertical motion problem.

VERTICAL MOTION TABLE

Variable	Value
v_0	0 m/s
v	?
Δy	2 m
a	9.8 m/s²
t	FIND

$$\Delta y = v_0 t + \tfrac{1}{2}at^2$$

$$\Delta y = \tfrac{1}{2}at^2$$

$$t = \sqrt{\frac{2\Delta y}{a}}$$

$$t = \sqrt{\frac{2(2m)}{9.8\,^m/_{s^2}}} = 0.639s$$

Now that you know the ball is in the air for 0.639 seconds, you can find how far it travels horizontally before reaching the ground. This is a horizontal motion problem, in which the acceleration is zero (nothing is causing the ball to accelerate horizontally). Because the ball doesn't accelerate, its initial velocity is also its final velocity, which is equal to its average velocity.

HORIZONTAL MOTION TABLE

Variable	Value
v_0	42 m/s
v	42 m/s
Δx	FIND
a	0 m/s²
t	0.639 s

$$\bar{v} = \frac{\Delta x}{t}$$

$$\Delta x = \bar{v}t = (42\,^m/_s)(0.639s) = 26.8m$$

You can therefore conclude that the baseball travels 26.8 meters horizontally before reaching the ground.

3.48 Q: The diagram below represents the path of a stunt car that is driv-
en off a cliff, neglecting friction.

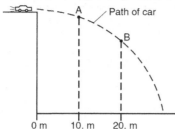

Compared to the horizontal component of the car's velocity at
point A, the horizontal component of the car's velocity at B is

(A) smaller

(B) greater

(C) the same

3.48 A: (C) the same. The car's horizontal acceleration is zero; therefore,
the horizontal velocity remains constant.

3.49 Q: A 0.2-kilogram red ball is thrown horizontally at a speed of 4 meters per second from a height of 3 meters. A 0.4-kilogram green
ball is thrown horizontally from the same height at a speed of 8
meters per second. Compared to the time it takes the red ball
to reach the ground, the time it takes the green ball to reach the
ground is

(A) one-half as great

(B) twice as great

(C) the same

(D) four times as great

3.49 A: (C) the same.   Both objects are thrown horizontally from the
same height. Because horizontal motion and vertical motion are
independent, both objects have the same vertical motion (they
both start with an initial vertical velocity of 0 m/s, have the same
acceleration of 9.8 m/s² down, and both travel the same vertical
distance). Therefore, the two objects reach the ground in the
same amount of time.

3.50 Q: A ball is thrown horizontally at a speed of 24 meters per second
from the top of a cliff. If the ball hits the ground 4.0 seconds later, approximately how high is the cliff?

(A) 6.0 m

(B) 39 m

(C) 78 m

(D) 96 m

3.50 A: (C) 78m

$$\Delta y = v_0 t + \tfrac{1}{2}at^2$$

$$\Delta y = \tfrac{1}{2}at^2$$

$$\Delta y = \tfrac{1}{2}(9.8\,^m\!/_{s^2})(4s)^2$$

$$\Delta y = 78m$$

3.51 Q: Projectile A is launched horizontally at a speed of 20 meters per second from the top of a cliff and strikes a level surface below, 3.0 seconds later. Projectile B is launched horizontally from the same location at a speed of 30 meters per second. The time it takes projectile B to reach the level surface is

(A) 4.5 s

(B) 2.0 s

(C) 3.0 s

(D) 10 s

3.51 A: (C) 3.0 s. They both take the same time to reach the ground because they both travel the same distance vertically, and they both have the same vertical acceleration (9.8 m/s² down) and initial vertical velocity (zero).

Angled Projectiles

For objects launched at an angle, you have to do a little more work to determine the initial velocity in both the horizontal and vertical directions. For example, if a football is kicked with an initial velocity of 40 m/s at an angle of 30° above the horizontal, you need to break the initial velocity vector up into x- and y-components in the same manner as covered in the components of vectors math review section.

Then, use the components for your initial velocities in your horizontal and vertical tables. Finally, don't forget that symmetry of motion also applies to the parabola of projectile motion. For objects launched and landing at the same height, the launch angle is equal to the landing angle. The launch velocity is equal to the landing velocity. And if you want an object to travel the maximum possible horizontal distance (or range), launch it at an angle of 45°.

3.52 Q: Herman the Human Cannonball is launched from level ground at an angle of 30° above the horizontal with an initial velocity of 26 m/s. How far does Herman travel horizontally before reuniting with the ground?

3.52 A: The first step in solving this type of problem is to determine Herman's initial horizontal and vertical velocity. You do this by breaking up his initial velocity into vertical and horizontal components:

$$v_{0_x} = v_0 \cos(\theta) = (26\,^m\!/_s)\cos(30°) = 22.5\,^m\!/_s$$

$$v_{0_y} = v_0 \sin(\theta) = (26\,^m\!/_s)\sin(30°) = 13\,^m\!/_s$$

Next, analyze Herman's vertical motion to find out how long he is in the air. You can analyze his motion on the way up, find the time, and double that to find his total time in the air:

VERTICAL MOTION TABLE

Variable	Value
v_0	13 m/s
v	0 m/s
Δy	?
a	-9.8 m/s²
t	FIND

$$v = v_0 + at$$

$$t = \frac{v - v_0}{a}$$

$$t_{up} = \frac{(0 - 13\,^m\!/_s)}{-9.8\,^m\!/_{s^2}} = 1.33s$$

$$t_{total} = 2 \times t_{up} = 2.65s$$

Now that you know Herman was in the air 2.65s, you can find how far he moved horizontally, using his initial horizontal velocity of 22.5 m/s.

HORIZONTAL MOTION TABLE

Variable	Value
v_0	22.5 m/s
v	22.5 m/s
Δx	FIND
a	0
t	2.65 s

$$\overline{v} = \frac{\Delta x}{t}$$

$$\Delta x = \overline{v}t$$

$$\Delta x = (22.5\,^m\!/_s)(2.65s) = 59.6m$$

Therefore, Herman must have traveled 59.6 meters horizontally before returning to the Earth.

3.53 Q: A child kicks a ball with an initial velocity of 8.5 meters per second at an angle of 35° with the horizontal, as shown. The ball has an initial vertical velocity of 4.9 meters per second and a total time of flight of 1.0 second. The maximum height reached by the ball is approximately: [Neglect air resistance]

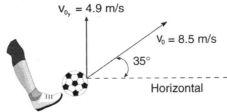

(A) 1.2 m

(B) 2.5 m

(C) 4.9 m

(D) 8.5 m

3.53 A: The maximum height in the air is a vertical motion problem. Start by recognizing that at its maximum height, the ball's vertical velocity will be zero, and it will have been in the air 0.5 seconds.

VERTICAL MOTION TABLE

Variable	Value
v_0	4.9 m/s
v	0 m/s
Δy	FIND
a	-9.8 m/s²
t	0.5 s

$$v^2 = v_0^2 + 2a\Delta y$$

$$\Delta y = \frac{v^2 - v_0^2}{2a}$$

$$\Delta y = \frac{(0\,{}^m\!/_s)^2 - (4.9\,{}^m\!/_s)^2}{2(-9.8\,{}^m\!/_{s^2})} = 1.2m$$

The correct answer must be (A) 1.2 meters. Note that you could have solved for the correct answer using any of the kinematic equations containing distance.

3.54 Q: A ball is thrown at an angle of 38° to the horizontal. What happens to the magnitude of the ball's vertical acceleration during the total time interval that the ball is in the air?

(A) It decreases, then increases.

(B) It decreases, then remains the same.

(C) It increases, then decreases.

(D) It remains the same.

3.54 A: (D) It remains the same since the acceleration of any projectile on the surface of Earth is 9.8 m/s² down the entire time the projectile is in the air.

3.55 Q: A golf ball is hit at an angle of 45° above the horizontal. What is the acceleration of the golf ball at the highest point in its trajectory? [Neglect friction.]

(A) 9.8 m/s² upward

(B) 9.8 m/s² downward

(C) 6.9 m/s² horizontal

(D) 0 m/s².

3.55 A: (B) 9.8 m/s² downward.

3.56 Q: A machine launches a tennis ball at an angle of 25° above the horizontal at a speed of 14 meters per second. The ball returns to level ground. Which combination of changes must produce an increase in time of flight of a second launch?

(A) decrease the launch angle and decrease the ball's initial speed

(B) decrease the launch angle and increase the ball's initial speed

(C) increase the launch angle and decrease the ball's initial speed

(D) increase the launch angle and increase the ball's initial speed

3.56 A: (D) will increase the ball's initial vertical velocity and therefore give the ball a larger time of flight. Increasing the launch angle and decreasing the ball's initial speed (answer B) COULD be a correct answer, depending on the magnitude of those changes, but the question asks which change MUST produce an increase in time of flight.

3.57 Q: A golf ball is given an initial speed of 20 meters per second and returns to level ground. Which launch angle above level ground results in the ball traveling the greatest horizontal distance? [Neglect friction.]

(A) 60°

(B) 45°

(C) 30°

(D) 15°

3.57 A: (B) 45° provides the greatest range for a projectile launched from level ground onto level ground, neglecting friction.

3.58 Q: A 30° incline sits on a 1.1-meter high table. A ball rolls off the incline with a velocity of 2 m/s. How far does the ball travel across the room before reaching the floor?

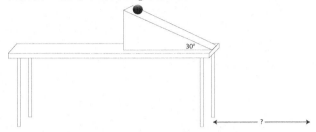

3.58 A: In order to determine how far the ball travels horizontally, you must first determine how long the ball is in the air. Begin by breaking up its initial velocity into horizontal and vertical components.

$$v_{0_x} = v_0 \cos(\theta) = (2.00\,{}^m\!/\!_s)\cos(30°) = 1.73\,{}^m\!/\!_s$$

$$v_{0_y} = v_0 \sin(\theta) = (2.00\,{}^m\!/\!_s)\sin(30°) = 1.00\,{}^m\!/\!_s$$

Determining how long the ball is in the air is a vertical motion problem. If you call the down direction positive, you can set up a vertical motion table as shown below.

VERTICAL MOTION TABLE

Variable	Value
v_0	1 m/s
v	?
Δy	1.1 m
a	9.8 m/s²
t	FIND

To solve this problem, you could solve for time in the kinematic equation: $\Delta y = v_0 t + \frac{1}{2}at^2$.

However, in doing so, you'll encounter a quadratic equation that will require you to utilize the quadratic formula. This is perfectly solvable, but you can save yourself some time and mathematical complexity if instead you solve for final velocity first, then solve for time.

$$v^2 = v_0^2 + 2a\Delta y$$

$$v^2 = (1\,{}^m\!/\!_s)^2 + 2(9.8\,{}^m\!/\!_{s^2})(1.1m)$$

$$v^2 = 22.56\,{}^{m^2}\!/\!_{s^2}$$

$$v = \sqrt{22.56\,{}^{m^2}\!/\!_{s^2}} = 4.75\,{}^m\!/\!_s$$

Now, knowing the ball's final vertical velocity, you can solve for the time the ball is in the air.

$$v = v_0 + at$$

$$t = \frac{v - v_0}{a}$$

$$t = \frac{4.75\,m/_s - 1\,m/_s}{9.8\,m/_{s^2}} = 0.383s$$

Next, knowing the time the ball is in the air, you can analyze the horizontal motion of the ball to calculate the horizontal distance traveled. Since you're neglecting air resistance, the horizontal acceleration of the ball is zero; therefore the initial velocity is the same as the final velocity.

HORIZONTAL MOTION TABLE

Variable	Value
v_0	1.73 m/s
v	1.73 m/s
Δx	FIND
a	0
t	0.383 s

$$\bar{v} = \frac{\Delta x}{t} \rightarrow \Delta x = \bar{v}t = (1.73\,m/_s)(0.383s) = 0.66m$$

The ball travels 0.66m across the room horizontally from the edge of the table before striking the ground.

Relative Velocity

You've probably heard the saying "motion is relative." Or perhaps you've heard people speak about Einstein's Theory of General Relativity and Einstein's Theory of Special Relativity. But what is this relativity concept?

In short, the concept of relative motion or relative velocity is all about understanding frame of reference. A frame of reference can be thought of as the state of motion of the observer of some event. For example, if you're sitting on a lawn chair watching a train travel past you from left to right at 50 m/s, you would consider yourself in a stationary frame of reference. From your perspective, you are at rest, and the train is moving. Further, assuming you have tremendous eyesight, you could even watch a glass of water sitting on a table inside the train move from left to right at 50 m/s.

An observer on the train itself, however, sitting beside the table with the glass of water, would view the glass of water as remaining stationary from their frame of reference. Because that observer is moving at 50 m/s, and the glass of water is moving at 50 m/s, the observer on the train sees no motion for the cup of water.

This seems like a simple and obvious example, yet when you take a step back and examine the bigger picture, you quickly find that all motion is relative. Going back to our original scenario, if you're sitting on your lawnchair watching a train go by, you believe you're in a stationary reference frame. The observer on the train looking out the window at you, however, sees you moving from right to left at 50 m/s.

Even more intriguing, an observer outside the Earth's atmosphere traveling with the Earth could use a "magic telescope" to observe you sitting in your lawnchair moving hundreds of meters per second as the Earth rotates about its axis. If this observer were further away from the Earth, he or she would also observe the Earth moving around the sun at speeds approaching 30,000 m/s. If the observer were even further away, they would observe the solar system (with the Earth, and you, on your lawnchair) orbiting the center of the Milky Way Galaxy at speeds approaching 220,000 m/s. And it goes on and on.

According to the laws of physics, there is no way to distinguish between an object at rest and an object moving at a constant velocity in an inertial (non-accelerating) reference frame. This means that there really is no "correct answer" to the question "how fast is the glass of water on the train moving?" You would be correct stating the glass is moving 50 m/s to the right and also correct in stating the glass is stationary. Imagine you're on a very smooth airplane, with all the window shades pulled down. It is physically impossible to determine whether you're flying through the air at a constant 300 m/s or whether you're sitting still on the runway. Even if you peeked out the window, you still couldn't say whether the plane was moving forward at 300 m/s, or the Earth was moving underneath the plane at 300 m/s.

As you observe (pun intended), how fast you are moving depends upon the observer's frame of reference. This is what is meant by the statement "motion is relative." In order to determine an object's velocity, you really need to also state the reference frame (i.e. the train moves 50 m/s with respect to the ground; the glass of water moves 50 m/s with respect to the ground; the glass of water is stationary with respect to the train.).

In most instances, the Earth makes a terrific frame of reference for physics problems. However, there are times when calculating the velocity of an object relative to different reference frames can be useful. Imagine you're in a canoe race, traveling down a river. It could be important to know not only your speed with respect to the flow of the river, but also your speed with respect to the riverbank, and even your speed with respect to your opponent's canoe in the race.

In dealing with these situations, you can state the velocity of an object with respect to its reference frame. For example, the velocity of object A with respect to reference frame C would be written as v_{AC}. Even if you don't know the velocity of object A with respect to C directly, by finding the velocity of object A with respect to some intermediate object B, and the velocity of object B with respect to C, you can combine your velocities using vector addition to obtain:

$$v_{AC} = v_{AB} + v_{BC}$$

This sounds more complicated than it actually is. Let's look at how this is applied in a few examples.

3.59 Q: A train travels at 60 m/s to the east with respect to the ground. A businessman on the train runs at 5 m/s to the west with respect to the train. Find the velocity of the man with respect to the ground.

3.59 A: First determine what information you are given. Calling east the positive direction, you know the velocity of the train with respect to the ground (v_{TG}=60 m/s). You also know the velocity of the man with respect to the train (v_{MT}=-5 m/s). Putting these together, you can find the velocity of the man with respect to the ground.

$$v_{MG} = v_{MT} + v_{TG} = -5 \tfrac{m}{s} + 60 \tfrac{m}{s} = 55 \tfrac{m}{s}$$

3.60 Q: An airplane flies at 250 m/s to the east with respect to the air. The air is moving at 15 m/s to the east with respect to the ground. Find the velocity of the plane with respect to the ground.

3.60 A: Again, start with the information you are given. If you call east positive, the velocity of the plane with respect to the air (v_{PA}) is 250 m/s. The velocity of the air with respect to the ground (v_{AG}) is 15 m/s. Solve for v_{PG}.

$$v_{PG} = v_{PA} + v_{AG} = 250 \tfrac{m}{s} + 15 \tfrac{m}{s} = 265 \tfrac{m}{s}$$

This strategy isn't limited to one-dimensional problems. Treating velocities as vectors, you can use vector addition to solve problems in multiple dimensions.

3.61 Q: The president's airplane, Air Force One, flies at 250 m/s to the east with respect to the air. The air is moving at 35 m/s to the north with respect to the ground. Find the velocity of Air Force One with respect to the ground.

3.61 A: In this case, it's important to realize that both v_{PA} and v_{AG} are two-dimensional vectors. Once again, you can find v_{PG} by vector addition.

$$v_{PG} = v_{PA} + v_{AG}$$

Drawing a diagram can be of tremendous assistance in solving this problem.

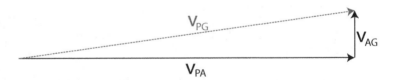

Looking at the diagram, you can easily solve for the magnitude of the velocity of the plane with respect to the ground using the Pythagorean Theorem.

$$v_{PG}^2 = v_{PA}^2 + v_{AG}^2 \qquad v_{PG} = \sqrt{v_{PA}^2 + v_{AG}^2}$$
$$v_{PG} = \sqrt{(250\,{}^m\!/_s)^2 + (35\,{}^m\!/_s)^2} = 252\,{}^m\!/_s$$

You can find the angle of Air Force One using basic trig functions.

$$\tan\theta = \frac{opp}{adj} = \frac{v_{AG}}{v_{PA}} \rightarrow \theta = \tan^{-1}\left(\frac{v_{AG}}{v_{PA}}\right) \rightarrow$$

$$\theta = \tan^{-1}\left(\frac{35\,^m\!/_s}{250\,^m\!/_s}\right) = 8°$$

Therefore, the velocity of Air Force One with respect to the ground is 252 m/s at an angle of 8° north of east.

Center of Mass

The motion of real objects is considerably more complex than that of simple theoretical particles. As you model the motion of complex objects, you are really analyzing the motion of a system made up of collections of substructures, all in motion as pieces of a whole. You are able to do this by treating the system as if it were a point particle with its entire mass located at a specific location known as the object's **center of mass**.

For symmetric objects, it is typically easy to determine the center of mass of the object by inspection. For example, the center of mass of a uniform sphere is at the very center of the sphere. For more complex objects, such as an armadillo, more involved mathematical or empirical techniques are required to find the center of mass. Mathematically, the center of mass of an object is the weighted average of the location of mass in an object.

For students taking the AP Physics 1 Exam only, AP-1 tests only a qualitative understanding of the center of mass. Students taking AP Physics 2, however, are expected to know how to calculate the center of mass of two-dimensional systems as treated below.

You can find the center of mass of a system of particles by taking the sum of the mass of the particles, multiplied by their positions, and dividing that by the total mass of the object. Looking at this in two dimensions, the center of mass in the x- and y-directions would be:

$$x_{cm} = \frac{\sum m_i x_i}{\sum m_i} = \frac{m_1 x_1 + m_2 x_2 + ...}{m_1 + m_2 + ...}$$

$$y_{cm} = \frac{\sum m_i y_i}{\sum m_i} = \frac{m_1 y_1 + m_2 y_2 + ...}{m_1 + m_2 + ...}$$

No matter how complex an object may be, you can calculate its center of mass and then treat the object as a point particle with total mass M. This allows you to apply basic physics principles to complex objects without adding unnecessary mathematical complexity to the analysis!

3.62 Q: Find the center of mass of an object modeled as two separate masses on the x-axis. The first mass is 2 kg at an x-coordinate of 2 and the second mass is 6 kg at an x-coordinate of 8.

3.62 A:
$$x_{cm} = \frac{\sum m_i x_i}{\sum m_i} = \frac{m_1 x_1 + m_2 x_2 + ...}{m_1 + m_2 + ...}$$

$$x_{cm} = \frac{(2kg)(2) + (6kg)(8)}{(2kg + 6kg)} = 6.5$$

This means you can treat the object as a point particle with a mass of 8 kg at an x-coordinate of 6.5 as shown below. Note that we are performing all calculations from the perspective of the origin as the reference point, though you could use any reference point you prefer.

Use the same strategy for finding the center of mass of a multi-dimensional object.

3.63 Q: Find the coordinates of the center of mass for the system shown below.

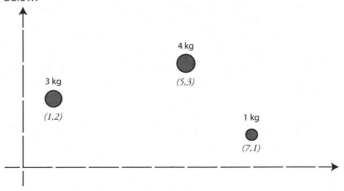

3.63 A:

$$x_{cm} = \frac{\sum m_i x_i}{\sum m_i} = \frac{(3kg)(1) + (4kg)(5) + (1kg)(7)}{(3kg + 4kg + 1kg)} = 3.75$$

$$y_{cm} = \frac{\sum m_i y_i}{\sum m_i} = \frac{(3kg)(2) + (4kg)(3) + (1kg)(1)}{(3kg + 4kg + 1kg)} = 2.38$$

Therefore the center of mass is a point particle with mass 8 kg located at (3.75, 2.38).

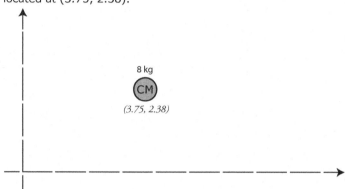

Test Your Understanding

1. Can an object have a negative acceleration and speed up? Explain.

2. Can an object accelerate at constant speed? Explain.

3. Design a procedure to measure your vertical leap using only a stopwatch. Try it. Compare your results to your vertical leap as measured with a meterstick. Which do you think is more accurate? Why?

4. Design an experiment to measure the acceleration due to gravity that uses your kinematic equations.

5. An object is dropped from a given height h at the same time an identical object is launched vertically upward from ground level. Determine the initial velocity of the object launched upward such that the two objects collide at height h/2.

6. Identify the center of mass of the following objects: a can of soda; a ruler; a protractor; a book; a baseball; and a horseshoe.

Chapter 4: Dynamics

"If I have seen further than others, it is by standing upon the shoulders of giants."

— Sir Isaac Newton

Objectives

1. Define mass and inertia and explain the meaning of Newton's 1st Law.
2. Define a force as an interaction between two objects and identify corresponding force pairs.
3. Utilize free body diagrams (FBDs) to identify the forces exerted on an object.
4. Resolve vectors into perpendicular components to create pseudo-FBDs.
5. Write and solve Newton's 2nd Law equations corresponding to given FBDs or pseudo-FBDs.
6. Predict the motion of an object due to multiple forces by applying Newton's 2nd Law of Motion.
7. Define friction and distinguish between static and kinetic friction.
8. Determine the coefficient of friction for two surfaces.
9. Calculate parallel and perpendicular components of an object's weight to solve ramp problems.
10. Analyze and solve basic Atwood Machine problems using Newton's 2nd Law of Motion.

Now that you've studied kinematics, you should have a pretty good understanding that objects in motion have **kinetic energy**, which is the ability of a moving object to move another object. To change an object's motion, and therefore its kinetic energy, the object must undergo a change in velocity, which is called an **acceleration**. So then, what causes an acceleration? To answer that question, you must study forces and their application.

Dynamics, or the study of forces, was very simply and effectively described by Sir Isaac Newton in 1686 in his masterpiece <u>Principia Mathematica Philosophiae Naturalis</u>. Newton described the relationship between forces and motion using three basic principles. Known as Newton's Laws of Motion, these concepts are still used today in applications ranging from sports science to aeronautical engineering.

Newton's 1st Law of Motion

Newton's 1st Law of Motion, also known as the **law of inertia**, can be summarized as follows:

> "An object at rest will remain at rest, and an object in motion will remain in motion, at constant velocity and in a straight line, unless acted upon by a net force."

This means that unless there is a net (unbalanced) force on an object, an object will continue in its current state of motion with a constant velocity. If this velocity is zero (the object is at rest), the object will continue to remain at rest. If this velocity is not zero, the object will continue to move in a straight line at the same speed. However, if a net (unbalanced) force does act on an object, that object's velocity will be changed (it will accelerate).

This sounds like a simple concept, but it can be quite confusing because it is difficult to observe this in everyday life. People are usually fine with understanding the first part of the law: "an object at rest will remain at rest unless acted upon by a net force." This is easily observable. The donut sitting on your breakfast table this morning didn't spontaneously accelerate up into the sky. Nor did the family cat, Whiskers, lounging sleepily on the couch cushion the previous evening, all of a sudden accelerate sideways off the couch for no apparent reason.

The second part of the law contributes a considerably bigger challenge to the conceptual understanding of this principle. Realizing that "an object in motion will continue in its current state of motion with constant velocity unless acted upon by a net force" isn't easy to observe here on Earth, making this law rather tricky. Almost all objects observed in everyday life that are in motion are being acted upon by a net force - friction. Try this example: take your physics book and give it a good push along the floor. As expected, the book moves for some distance, but rather rapidly slides to a halt. An outside force, friction, has acted upon it. Therefore, from typical observations, it would be easy to think that an object must have a force continually applied upon it to remain in motion. However, this isn't so. If you took the same book out into the far reaches of space, away from any gravitational or frictional forces, and pushed it away, it would continue moving in a straight line at a constant velocity forever, as there are no external forces to change its motion. When the net force on an object is 0, the object is in **static equilibrium**. You'll revisit static equilibrium when discussing Newton's 2nd Law.

The tendency of an object to resist a change in velocity is known as the object's **inertia**. For example, a train has significantly more inertia than a skateboard. It is much harder to change the train's velocity than it is the skateboard's. The measure of an object's inertia is its **inertial mass**, typically referred to as mass. In other words, the more mass an object has, the greater its inertia.

4.1 Q: A 0.50-kilogram cart is rolling at a speed of 0.40 meter per second. If the speed of the cart is doubled, the inertia of the cart is

(A) halved

(B) doubled

(C) quadrupled

(D) unchanged

4.1 A: (D) unchanged. Mass is a measure of an object's inertia, and the mass of the cart is constant.

4.2 Q: Which object has the greatest inertia?

(A) a falling leaf

(B) a softball in flight

(C) a seated high school student

(D) a rising helium-filled toy balloon

4.2 A: (C) a seated high school student has the greatest inertia (mass).

4.3 Q: Which object has the greatest inertia?

(A) a 5.00-kg mass moving at 10.0 m/s

(B) a 10.0-kg mass moving at 1.00 m/s

(C) a 15.0-kg mass moving at 10.0 m/s

(D) a 20.0-kg mass moving at 1.00 m/s

4.3 A: (D) a 20.0-kg mass has the greatest inertia.

If you recall from the kinematics unit, a change in velocity is known as an acceleration. Therefore, the second part of this law could be re-written to state that an object acted upon by a net force will be accelerated.

But what exactly is a force? A **force** is a vector quantity describing the push or pull on an object. Forces are measured in newtons (N), named after Sir Isaac Newton, of course. A newton is not a base unit, but is instead a derived unit, equivalent to $1 \text{ kg} \times \text{m/s}^2$. Interestingly, the gravitational force on a medium-sized apple is approximately 1 newton.

You can break forces down into two basic types: contact forces and field forces. Contact forces occur when objects touch each other. Examples of contact forces include tension, friction, normal forces, elastic forces such as springs, and even the buoyant force. Field forces, also known as non-contact forces, occur at a distance. Examples of field forces include the gravitational force, the magnetic force, and the electrical force between two charged objects.

Interestingly, as you examine the universe on extremely small scales and on the most basic level, contact forces actually arise from interatomic electric field forces between objects. Therefore, viewed from a very basic perspective, all forces are ultimately field forces.

So, what then is a net force? A **net force** is just the vector sum of all the forces acting on an object. Imagine you and your sister are fighting over the last Christmas gift. You are pulling one end of the gift toward you with a force of 5N. Your sister is pulling the other end toward her (in the opposite direction) with a force of 5N. The net force on the gift, then, would be 0N; therefore, there would be no net force. As it turns out, though, you have a passion for Christmas gifts, and now increase your pulling force to 6N. The net force on the gift now is 1N in your direction; therefore, the gift would begin to accelerate toward you (yippee!). It can be difficult to keep track of all the forces acting on an object.

Free Body Diagrams

Fortunately, we have a terrific tool for analyzing the forces acting upon objects. This tool is known as a free body diagram. Quite simply, a **free body diagram** is a representation of a single object, or system, with vector arrows showing all the external forces acting on the object. These diagrams

Chapter 4: Dynamics

make it very easy to identify exactly what the net force is on an object, and they're also quite simple to create:

1. Isolate the object of interest. Draw the object as a point particle representing the same mass.
2. Sketch and label each of the external forces acting on the object.
3. Choose a coordinate system, with the direction of motion as one of the positive coordinate axes.
4. If all forces do not line up with your axes, resolve those forces into components using trigonometry (note that the formulas below only work if the angle is measured from the horizontal).

$$A_x = A\cos(\theta)$$
$$A_y = A\sin(\theta)$$

5. Redraw your free body diagram, replacing forces that don't overlap the coordinates axes with their components.

As an example, picture a glass of soda sitting on the dining room table. You can represent the glass of soda in the diagram as a single dot. Then, represent each of the vector forces acting on the soda by drawing arrows and labeling them. In this case, you can start by recognizing the force of gravity on the soda, known more commonly as the soda's **weight**. Although you could label this force as F_{grav}, or W, get in the habit right now of writing the force of gravity on an object as mg. You can do this because the force of gravity on an object is equal to the object's mass times the acceleration due to gravity, g.

Of course, since the soda isn't accelerating, there must be another force acting on the soda to balance out the weight. This force, the force of the table pushing up on the soda, is known as the **normal force** (F_N). In physics, the normal force refers to a force perpendicular to a surface (normal in this case meaning perpendicular). The force of gravity on the soda must exactly match the normal force on the soda, although they are in opposite directions; therefore, there is no net force on the soda. The free body diagram for this situation could be drawn as shown at left.

4.4 Q: Which diagram represents a box in equilibrium?

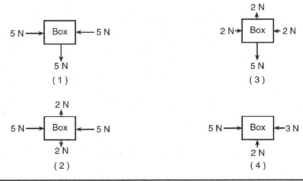

4.4 A: (2) all forces are balanced for a net force of zero.

4.5 Q: If the sum of all the forces acting on a moving object is zero, the
                object will
                (A) slow down and stop
                (B) change the direction of its motion
                (C) accelerate uniformly
                (D) continue moving with constant velocity

4.5 A: (D) continue moving with constant velocity per Newton's 1st Law.

Newton's 2nd Law of Motion

Newton's 2nd Law of Motion may be the most important principle in all of
modern-day physics because it explains exactly how an object's velocity is
changed by a net force. In words, Newton's 2nd Law states that "the acceleration of an object is directly proportional to the net force applied, and
inversely proportional to the object's mass." In equation form:

$$\vec{a} = \frac{\sum \vec{F}}{m} = \frac{\vec{F}_{net}}{m}$$

It's important to remember that both force and acceleration are vectors.
Therefore, the direction of the acceleration, or the change in center-of-
mass velocity of an object, will be in the same direction as the net force.
You can also look at this equation from the opposite perspective. A net
force applied to an object changes an object's velocity (produces an acceleration), and is frequently written as:

$$\vec{F}_{net} = m\vec{a}$$

You can analyze many situations involving both balanced and unbalanced
forces on an object using the same basic steps.

1. Draw a free body diagram.
2. For any forces that don't line up with the x- or y-axes, break
    those forces up into components that do lie on the x- or y-axis.
3. Write expressions for the net force in x- and y- directions. Set
    the net force equal to ma, since Newton's 2nd Law tells us that
    F=ma.
4. Solve the resulting equations.

Let's take a look and see how these steps can be applied to a sample problem.

4.6 Q: A force of 25 newtons east and a force of 25 newtons west act concurrently on a 5-kilogram cart. Find the acceleration of the cart.

(A) 1.0 m/s² west

(B) 0.20 m/s² east

(C) 5.0 m/s² east

(D) 0 m/s²

4.6 A: Step 1: Draw a free-body diagram (FBD).

Step 2: All forces line up with x-axis. Define east as positive.

Step 3: $F_{net} = 25N - 25N = ma$

Step 4: $0 = ma$

$a = 0$

Correct answer must be (D) 0 m/s².

Of course, everything you've already learned about kinematics still applies, and can be applied to dynamics problems as well.

4.7 Q: A 0.15-kilogram baseball moving at 20 m/s is stopped by a catcher in 0.010 seconds. The average force stopping the ball is

(A) 3.0×10⁻² N

(B) 3.0×10⁰ N

(C) 3.0×10¹ N

(D) 3.0×10² N

4.7 A: First write down what information is given and what we're asked to find. Define the initial direction of the baseball as positive.

Given: Find:

$m = 0.15 kg$ F

$v_0 = 20\,{}^m\!/_s$

$v = 0\,{}^m\!/_s$

$t = 0.010 s$

Use kinematics to find acceleration.

$$a = \frac{\Delta v}{t} = \frac{v - v_0}{t} \rightarrow$$

$$a = \frac{0\,{}^m\!/_s - 20\,{}^m\!/_s}{0.010 s} = -2000\,{}^m\!/_{s^2}$$

The negative acceleration indicates the acceleration is in the direction opposite that of the initial velocity of the baseball. Now that you know acceleration, you can solve for force using Newton's 2nd Law.

$$F_{net} = ma = (0.15)(-2000 \,{}^{m}/_{s^2}) = -300N$$

The correct answer must be (D), 300 newtons. The negative sign in our answer indicates that the force applied is opposite the direction of the baseball's initial velocity.

4.8 Q: Two forces, F_1 and F_2, are applied to a block on a frictionless, horizontal surface as shown below.

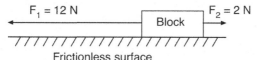

F₁ = 12 N   F₂ = 2 N

Block

Frictionless surface

If the magnitude of the block's acceleration is 2.0 meters per second², what is the mass of the block?

(A) 1 kg

(B) 5 kg

(C) 6 kg

(D) 7 kg

4.8 A: Define left as the positive direction.

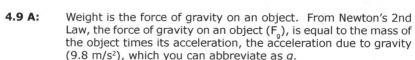

12 N   2N

$$F_{net} = 12N - 2N = 10N = ma$$

$$m = \frac{F_{net}}{a} = \frac{10N}{2 \,{}^{m}/_{s^2}} = 5kg$$

(B) 5 kg

4.9 Q: What is the weight of a 2.00-kilogram object on the surface of Earth?

(A) 4.91 N

(B) 2.00 N

(C) 9.81 N

(D) 19.6 N

4.9 A: Weight is the force of gravity on an object. From Newton's 2nd Law, the force of gravity on an object (F_g), is equal to the mass of the object times its acceleration, the acceleration due to gravity (9.8 m/s²), which you can abbreviate as g.

$$F_g = ma$$

$$W = mg$$

$$W = (2kg)(9.8 \, ^m/_{s^2}) = 19.6N$$

(D) 19.6 N is correct.

4.10 Q: A 25-newton horizontal force northward and a 35-newton horizontal force southward act concurrently on a 15-kilogram object on a frictionless surface. What is the magnitude of the object's acceleration?

(A) 0.67 m/s²

(B) 1.7 m/s²

(C) 2.3 m/s²

(D) 4.0 m/s²

4.10 A: (A) 0.67 m/s².

$$a = \frac{F_{net}}{m}$$

$$a = \frac{35N - 25N}{15kg} = 0.67 \, ^m/_{s^2}$$

4.11 Q: A cardboard box of mass m on a wooden floor is represented by the free body diagram below.

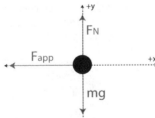

Given the FBD, which of the following expressions could be accurate mathematical representations for the box? Choose all that apply.

(A) F_N-mg=ma$_x$

(B) -F_{app} =ma$_x$

(C) F_N-mg=ma$_y$

(D) F_{app}=ma$_y$

4.11 A: (B) and (C): -F_{app} =ma$_x$ and F_N-mg=ma$_y$

4.12 Q: Three objects with differing masses are connected by strings and pulled by the right-most string with tension T_1 across a frictionless surface as shown in the diagram below.

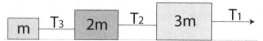

Which of the following expressions accurately depicts the acceleration of the system? Choose all that apply.

(A) T3/2m

(B) T2/3m

(C) T1/6m

(D) None of the above.

4.12 A: Both (B) and (C) are correct.

Writing Newton's 2nd Law equations in the x-direction for each of the three objects from left to right:

$$\vec{F}_{net_x} = m\vec{a}_x$$

$$\vec{T}_3 = m\vec{a} \rightarrow \vec{a} = \frac{\vec{T}_3}{m}$$

$$\vec{T}_2 - \vec{T}_3 = 2m\vec{a} \xrightarrow[\vec{T}_2 = 3m\vec{a}]{\vec{T}_3 = m\vec{a}} \vec{a} = \frac{\vec{T}_2}{3m}$$

$$\vec{T}_1 - \vec{T}_2 = 3m\vec{a} \xrightarrow[\vec{T}_1 = 6m\vec{a}]{\vec{T}_2 = 3m\vec{a}} \vec{a} = \frac{\vec{T}_1}{6m}$$

The special situation in which the net force on an object turns out to be zero, called **static equilibrium**, tells you immediately that the object isn't accelerating. If the object is moving with some velocity, it will remain moving with that exact same velocity. If the object is at rest, it will remain at rest. Sounds familiar, doesn't it? This is a restatement of Newton's 1st Law of Motion, the Law of Inertia. So in reality, Newton's 1st Law is just a special case of Newton's 2nd Law, describing static equilibrium conditions! Consider the situation of a tug-of-war... if both participants are pulling with tremendous force, but the force is balanced, there is no acceleration -- a great example of static equilibrium.

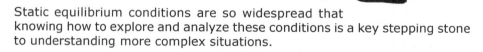

Static equilibrium conditions are so widespread that knowing how to explore and analyze these conditions is a key stepping stone to understanding more complex situations.

One common analysis question involves finding the equilibrant force given a free body diagram of an object. The **equilibrant** is a single force vector that you add to the unbalanced forces on an object in order to bring it into static equilibrium. For example, if you are given a force vector of 10N north and 10N east, and asked to find the equilibrant, you're really being asked to find a force that will offset the two given forces, bringing the object into static equilibrium.

To find the equilibrant, you must first find the net force being applied to the object. To do this, apply your vector math and add up the two vectors by first lining them up tip to tail, then drawing a straight line from the starting point of the first vector to the ending point of the last vector. The magnitude of this vector can be found using the Pythagorean Theorem.

Finally, to find the equilibrant vector, add a single vector to the diagram that will give a net force of zero. If your total net force is currently 14N northeast, then the vector that should bring this back into equilibrium, the equilibrant, must be the opposite of 14N northeast, or a vector with magnitude 14N to the southwest.

4.13 Q: A 20-newton force due north and a 20-newton force due east act concurrently on an object, as shown in the diagram below.

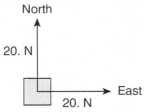

The additional force necessary to bring the object into a state of equilibrium is

(A) 20 N northeast

(B) 20 N southwest

(C) 28 N northeast

(D) 28 N southwest

4.13 A: (D) The resultant vector is 28 newtons northeast, so its equilibrant must be 28 newtons southwest

Another common analysis question involves asking whether three vectors could be arranged to provide a static equilibrium situation.

4.14 Q: A 3-newton force and a 4-newton force are acting concurrently on a point. Which force could not produce equilibrium with these two forces?

(A) 1 N

(B) 7 N

(C) 9 N

(D) 4 N

4.14 A: (C) A 9-newton force could not produce equilibrium with a 3-newton and a 4-newton force. One way to test this is to draw vectors of the three forces. If you can arrange the vectors to create a closed triangle, they can produce equilibrium.

4.15 Q: A net force of 10 newtons accelerates an object at 5.0 meters per second². What net force would be required to accelerate the same object at 1.0 meter per second²?

(A) 1.0 N

(B) 2.0 N

(C) 5.0 N

(D) 50 N

4.15 A: Strategy: First, solve for the mass of the object.

$$F_{net} = ma$$

$$m = \frac{F_{net}}{a} = \frac{10N}{5\,{}^m\!/_{s^2}} = 2kg$$

Next, use Newton's 2nd Law to determine the force required to accelerate the object at 1 m/s².

$$F_{net} = ma = (2kg)(1\,{}^m\!/_{s^2}) = 2N$$

The correct answer is (B) 2.0 N.

4.16 Q: A 1.0-newton metal disk rests on an index card that is balanced on top of a glass. What is the net force acting on the disk?

(A) 1 N

(B) 2 N

(C) 0 N

(D) 9.8 N

4.16 A: (C) 0N because the disk is at rest and isn't accelerating, so the net force must be zero.

4.17 Q: A 1200-kilogram space vehicle travels at 4.8 meters per second along the level surface of Mars. If the magnitude of the gravitational field strength on the surface of Mars is 3.7 newtons per kilogram, the magnitude of the normal force acting on the vehicle is

(A) 320 N

(B) 930 N

(C) 4400 N

(D) 5800 N

4.17 A: If the gravitational field strength is 3.7 N/kg, and the space vehicle weighs 1200 kg, the gravitational force on the space vehicle must be 1200kg × (3.7 N/kg) = 4440N. If there's a downward force of 4440N due to gravity, the normal force must be equal and opposite, or 4440N upward; therefore the best answer is (C) 4400N.

4.18 Q: Which body is in equilibrium?

(A) a satellite orbiting Earth in a circular orbit

(B) a ball falling freely toward the surface of Earth

(C) a car moving with constant speed along a straight, level road

(D) a projectile at the highest point in its trajectory

4.18 A: (C) a car moving with constant speed.

4.19 Q: A 15-kg wagon is pulled to the right across a surface by a tension of 100 newtons at an angle of 30 degrees above the horizontal. A frictional force of 20 newtons to the left acts simultaneously. What is the acceleration of the wagon?

4.19 A: First, draw a diagram of the situation.

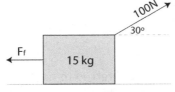

Next, create a FBD and pseudo-FBD.

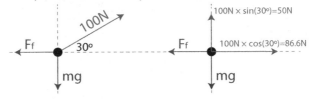

Finally, use Newton's 2nd Law in the x-direction to determine the horizontal acceleration of the wagon.

$$F_{net_x} = ma_x \rightarrow 86.6N - 20N = ma_x \rightarrow$$

$$a_x = \frac{66.6N}{15kg} = 4.44\,{}^m\!/_{s^2}$$

4.20 Q: A traffic light is suspended by two cables as shown in the diagram. If cable 1 has a tension T_1=49 newtons, and cable 2 has a tension T_2=85 newtons, find the mass of the traffic light.

4.20 A: First draw the FBD and pseudo-FBD for the traffic light.

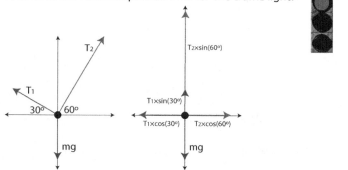

Now, you can write a Newton's 2nd Law Equation using the pseudo-FBD and solve for the mass of the stoplight.

$$F_{net_y} = T_1 \sin 30° + T_2 \sin 60° - mg = 0 \rightarrow$$

$$m = \frac{T_1 \sin 30° + T_2 \sin 60°}{g} = \frac{49N \sin 30° + 85N \sin 60°}{9.8\,{}^m\!/_{s^2}} \rightarrow$$

$$m = 10kg$$

Newton's 3rd Law

Newton's 3rd Law of Motion, commonly referred to as the Law of Action and Reaction, describes the phenomena by which all forces come in pairs. A force is really an interaction between two objects. If Object 1 exerts a force on Object 2, then Object 2 must exert a force back on Object 1, and the force of Object 1 on Object 2 is equal in magnitude, or size, but opposite in direction to the force of Object 2 on Object 1. Written mathematically:

$$\vec{F}_{1on2} = -\vec{F}_{2on1}$$

This has many implications, many of which aren't immediately obvious. For example, if you punch the wall with your fist with a force of 100N, the wall imparts a force back on your fist of 100N (which is why it hurts!). Or try this: push on the corner of your desk with your palm for a few seconds. Now look at your palm... see the indentation? That's because the corner of the desk pushed back on your palm.

Although this law surrounds your actions everyday, often times you may not even realize its effects. To run forward, a cat pushes with its legs backward on the ground, and the ground pushes the cat forward. How do you swim? If you want to swim forwards, which way do you push on the water? Backwards, that's right. As you push backwards on the water, the reactionary force, the water pushing you, propels you forward. How do you jump up in the air? You push down on the ground, and it's the reactionary force of the ground pushing on you that accelerates you skyward!

As you can see, then, forces always come in pairs because forces result from the interaction of two objects. These pairs are known as **action-reaction pairs**. With this in mind, you can see why an object can never exert a force upon itself; there must be another object to interact with in order to generate force pairs. What are the action-reaction force pairs for a girl kicking a soccer ball? The girl's foot applies a force on the ball, and the ball applies an equal and opposite force on the girl's foot. Further, if two objects interact with each other by applying a force upon each other, and the objects are part of the same system, the velocity of the system's center of mass cannot change.

How does a rocket ship maneuver in space? The rocket propels hot expanding gas particles outward, so the gas particles in return push the rocket forward. Newton's 3rd Law even applies to gravity. The Earth exerts a gravitational force on you (downward). You, therefore, must apply a gravitational force upward on the Earth!

4.21 Q: Earth's mass is approximately 81 times the mass of the Moon. If Earth exerts a gravitational force of magnitude F on the Moon, the magnitude of the gravitational force of the Moon on Earth is

(A) F

(B) F/81

(C) 9F

(D) 81F

4.21 A: (A) The force Earth exerts on the Moon is the same in magnitude and opposite in direction of the force the Moon exerts on Earth.

4.22 Q: A 400-newton girl standing on a dock exerts a force of 100 new-tons on a 10,000-newton sailboat as she pushes it away from the dock. How much force does the sailboat exert on the girl?

(A) 25 N

(B) 100 N

(C) 400 N

(D) 10,000 N

4.22 A: (B) The force the girl exerts on the sailboat is the same in magnitude and opposite in direction of the force the sailboat exerts on the girl.

4.23 Q: A carpenter hits a nail with a hammer. Com-pared to the magnitude of the force the hammer exerts on the nail, the magnitude of the force the nail exerts on the hammer during contact is

(A) less

(B) greater

(C) the same

4.23 A: (C) the same per Newton's 3rd Law.

Friction

Up until this point, it's been convenient to ignore one of the most useful and most troublesome forces in everyday life... a force that has tremendous application in transportation, machinery, and all parts of mechanics, yet people spend tremendous amounts of effort and money each day fighting it. This force, **friction**, is a force that opposes motion.

4.24 Q: A projectile launched at an angle of 45° above the horizontal trav-els through the air. Compared to the projectile's theoretical path with no air friction, the actual trajectory of the projectile with air friction is

(A) lower and shorter

(B) lower and longer

(C) higher and shorter

(D) higher and longer

4.24 A: (A) lower and shorter. Friction opposes motion.

4.25 Q: A box is pushed toward the right across a classroom floor. The force of friction on the box is directed toward the

(A) left

(B) right

(C) ceiling

(D) floor

4.25 A: (A) left. Friction opposes motion.

There are two main types of friction. **Kinetic friction** is a frictional force that opposes motion for an object which is sliding along another surface. **Static friction**, on the other hand, acts on an object that isn't sliding. If you push on your textbook, but not so hard that it slides along your desk, static friction is opposing your applied force on the book, leaving the book in static equilibrium.

The magnitude of the frictional force depends upon two factors:

1. The nature of the surfaces in contact.
2. The normal force acting on the object (F_N).

The ratio of the frictional force and the normal force provides the **coefficient of friction** (μ), a proportionality constant that is specific to the two materials in contact.

You can look up the coefficient of friction for various surfaces. Make sure you choose the appropriate coefficient. Use the static coefficient (μ_s) for objects which are not sliding, and the kinetic coefficient (μ_k) for objects which are sliding.

Approximate Coefficients of Friction		
	Kinetic	Static
Rubber on concrete (dry)	0.68	0.90
Rubber on concrete (wet)	0.58	
Rubber on asphalt (dry)	0.67	0.85
Rubber on asphalt (wet)	0.53	
Rubber on ice	0.15	
Waxed ski on snow	0.05	0.14
Wood on wood	0.30	0.42
Steel on steel	0.57	0.74
Copper on steel	0.36	0.53
Teflon on Teflon	0.04	

Which coefficient would you use for a sled sliding down a snowy hill? The kinetic coefficient of friction, of course. How about a refrigerator on your linoleum floor that is at rest and you want to start in motion? That would be the static coefficient of friction. Let's try a harder one: a car drives with its tires rolling freely. Is the friction between the tires and the road static or kinetic? Static. The tires are in constant contact with the road, much like walking. If the car was skidding, however, and the tires were locked, you would look at kinetic friction. Let's take a look at a sample problem:

4.26 Q: A car's performance is tested on various horizontal road surfaces. The brakes are applied, causing the rubber tires of the car to slide along the road without rolling. The tires encounter the greatest force of friction to stop the car on

(A) dry concrete

(B) dry asphalt

(C) wet concrete

(D) wet asphalt

4.26 A: To obtain the greatest force of friction (F_f), you'll need the greatest coefficient of friction (μ). Use the kinetic coefficient of friction (μ_k) since the tires are sliding. From the Approximate Coefficients of Friction table, the highest kinetic coefficient of friction for rubber comes from rubber on dry concrete. Answer: (A).

4.27 Q: The diagram below shows a block sliding down a plane inclined at angle θ with the horizontal.

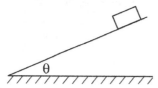

As angle θ is increased, the coefficient of kinetic friction between the bottom surface of the block and the surface of the incline will

(A) decrease

(B) increase

(C) remain the same

4.27 A: (C) remain the same. Coefficient of friction depends only upon the materials in contact.

The normal force always acts perpendicular to a surface, and comes from the interaction between atoms that act to maintain its shape. In many cases, it can be thought of as the elastic force trying to keep a flat surface flat (instead of bowed). You can use the normal force to calculate the magnitude of the frictional force.

The force of friction, depending only upon the nature of the surfaces in contact (μ) and the magnitude of the normal force (F_N), can be determined using the formula:

$$|F_f| \le \mu |F_N|$$

It is important to note that the frictional force will match the applied force until the applied force overcomes the maximum frictional force, which occurs when the magnitude of the frictional force is equal to the coefficient of friction multiplied by the magnitude of the normal force ($F_f = \mu F_N$).

Solving problems involving friction requires application of the same basic principles you've been learning about throughout the dynamics unit: drawing a free body diagram, applying Newton's 2nd Law along the x- and/or y-axes, and solving for any unknowns. The only new skill is drawing the frictional force on the free body diagram, and using the relationship between the force of friction and the normal force to solve for the unknowns. Let's take a look at another sample problem:

4.28 Q: The diagram below shows a 4.0-kilogram object accelerating at 10 meters per second² on a rough horizontal surface.

Acceleration = 10. m/s² ⟶

| Frictional force = F_f | m = 4.0 kg | Applied force = 50. N |

(Not drawn to scale)

What is the magnitude of the frictional force F_f acting on the object?

(A) 5.0 N

(B) 10 N

(C) 20 N

(D) 40 N

4.28 A: Define to the right as the positive direction.

$$F_{net} = ma$$

$$F_{app} - F_f = ma$$

$$F_f = F_{app} - ma$$

$$F_f = 50N - (4kg)(10 \, ^m\!/_{s^2}) = 10N$$

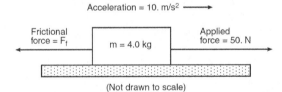

Answer: (B) 10 N.

Let's take a look at a more involved problem, tying together free body diagrams, Newton's 2nd Law, and the coefficient of friction:

4.29 Q: An ice skater applies a horizontal force to a 20-kilogram block on frictionless, level ice, causing the block to accelerate uniformly at 1.4 m/s² to the right. After the skater stops pushing the block, it slides onto a region of ice that is covered by a thin layer of sand. The coefficient of kinetic friction between the block and the sand-covered ice is 0.28. Calculate the magnitude of the force applied to the block by the skater.

4.29 A: Define right as the positive direction.

$$F_{net} = ma$$

$$F_{net} = (20kg)(1.4 \, \text{m}/_{s^2}) = 28N$$

4.30 Q: Referring to the previous problem, determine the magnitude of the normal force acting on the block.

4.30 A: $$F_{net_y} = ma_y$$

$$F_{net_y} = F_N - mg = 0$$

$$F_N = mg = (20kg)(9.8 \, \text{m}/_{s^2}) = 196N$$

4.31 Q: Referring to the previous ice skater problem, calculate the magnitude of the force of friction acting on the block as it slides over the sand-covered ice.

4.31 A: $$F_f = \mu F_N$$

$$F_f = (0.28)(196N) = 55N$$

Note that a new FBD is required when forces on the object change.

These same steps can be used in many different ways in many different problems, but the same basic problem solving methodology still works... draw a free body diagram, apply Newton's 2nd Law, utilize the friction formula if necessary, and solve!

Chapter 4: Dynamics

4.32 Q: A horizontal force of 8.0 newtons is used to pull a 20-newton wooden box moving toward the right along a horizontal, wood surface, as shown.

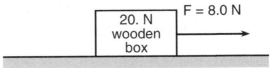

Wood

A) Calculate the magnitude of the frictional force acting on the box.

B) Determine the magnitude of the net force acting on the box.

C) Determine the mass of the box.

D) Calculate the magnitude of the acceleration of the box.

4.32 A: A) Recognize that the box has a weight of 20N, therefore its normal force must be 20N since it is not accelerating vertically.

$$F_f = \mu F_N = (0.30)(20N) = 6N$$

B) $F_{net} = F_{app} - F_f = 8N - 6N = 2N$

C) $mg = 20N$

$$m = \frac{20N}{g} = \frac{20N}{9.8\,m/_{s^2}} = 2.04kg$$

D) $a = \dfrac{F_{net}}{m} = \dfrac{2N}{2.04kg} = 0.98\,m/_{s^2}$

4.33 Q: Compared to the force needed to start sliding a crate across a rough level floor, the force needed to keep it sliding once it is moving is

(A) less

(B) greater

(C) the same

4.33 A: (A) less, since kinetic friction is less than static friction.

4.34 Q: An airplane is moving with a constant velocity in level flight. Compare the magnitude of the forward force provided by the engines to the magnitude of the backward frictional drag force.

4.34 A: The forces must be the same since the plane is moving with constant velocity.

4.35 Q: When a 12-newton horizontal force is applied to a box on a horizontal tabletop, the box remains at rest. The force of static friction acting on the box is

(A) 0 N

(B) between 0 N and 12 N

(C) 12 N

(D) greater than 12 N

4.35 A: (C) 12 N. Because the box is at rest in static equilibrium, all forces on it must be balanced, therefore the force of static friction must be 12N.

4.36 Q: Suzie pulls the handle of a 20-kilogram sled across the yard with a force of 100N at an angle of 30 degrees above the horizontal. The yard exerts a frictional force of 25N on the sled.

A) Find the coefficient of friction between the sled and the yard.

B) Determine the distance the sled travels if it starts from rest and Suzie maintains her 100N force for five seconds.

4.36 A: A) First draw a FBD and pseudo-FBD for the situation.

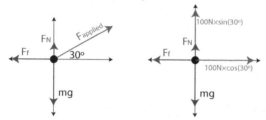

Next, write a Newton's 2nd Law Equation in the y-direction to solve for the normal force.

$$F_{net_y} = 100N\sin(30°) + F_N - mg = 0 \rightarrow$$

$$F_N = mg - 100N\sin(30°) = 200N - 50N = 150N$$

Next, find the coefficient of friction by dividing the frictional force by the normal force.

$$\mu = \frac{F_f}{F_N} = \frac{25N}{150N} = 0.167$$

B) In order to determine the distance the sled travels if it starts from rest, first find the acceleration of the sled by writing a Newton's 2nd Law Equation in the x-direction and solving for the horizontal acceleration.

$$F_{net_x} = 100N\cos(30°) - F_f = 86.6N - 25N = ma_x \rightarrow$$

$$a_x = \frac{61.6N}{20kg} = 3.1 \frac{m}{s^2}$$

Use the kinematic equations to find the distance the sled travels.

$$x = x_0 + v_{x0}t + \tfrac{1}{2}a_x t^2 = \tfrac{1}{2}(3.1 \tfrac{m}{s^2})(5s)^2 = 38.8m$$

4.37 Q: A 10-kg wooden box sits on a wooden surface. The coefficient of static friction between the two surfaces is 0.42. A horizontal force of 20N is applied to the box. What is the force of friction on the box?

4.37 A: Start by finding the maximum force of friction on the box.

$$\left|F_f\right| \le \mu \left|F_N\right| \rightarrow F_f \le (0.42)(10kg)(9.8 \tfrac{m}{s^2}) \rightarrow$$
$$F_f \le 41.2N$$

The force of friction will match the applied force on the box to keep it at rest and in static equilibrium as long as the applied force remains less than 41.2N. Therefore, the force of friction on the box must be 20N.

Air Resistance

Throughout this course, a common source of friction observed in everyday life is often neglected. Air resistance, also known as drag or fluid resistance, is a form of fluid friction between a moving object and the air around it, and can play a significant role as a force affecting moving bodies. Modeling air resistance mathematically can be a detailed and involved process because the force of air resistance depends on the object's velocity.

When an object falls through the atmosphere, it accelerates until it reaches its maximum vertical velocity, known as its **terminal velocity** (v_t). Once an object reaches its terminal velocity, it continues to fall at constant velocity with no acceleration. In this situation, the force of gravity on the object balances the drag force on the object. With zero net force, the object is in equilibrium, and continues to fall at a constant velocity as indicated by Newton's 1st Law of Motion.

Assume David falls from an airplane. Typically, the drag forces on a free-falling object take a form similar to $F_{drag}=bv$ or $F_{drag}=cv^2$, where b and c are constants. For the sake of this problem, let's assume $F_{drag}=bv$. You could then write a Newton's 2nd Law Equation in the y-direction for the skydiver.

$$F_{net_Y} = mg - F_{drag} = ma_Y \xrightarrow{F_{drag}=bv} mg - bv = ma_Y$$

Initially, at time t=0, velocity=0, therefore a=g. After a long time, however, David reaches terminal velocity (v_t) and a=0. At this point, F_{drag}=mg.

$$mg - bv = ma_Y \xrightarrow[v=v_t]{a_Y=0} mg - bv_t = 0 \rightarrow v_t = \frac{mg}{b}$$

A brief analysis of the acceleration-time, velocity-time, and displacement-time graphs can provide further insight into the situation.

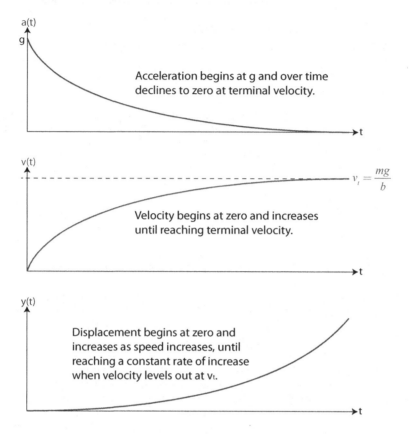

Acceleration begins at g and over time declines to zero at terminal velocity.

$$v_t = \frac{mg}{b}$$

Velocity begins at zero and increases until reaching terminal velocity.

Displacement begins at zero and increases as speed increases, until reaching a constant rate of increase when velocity levels out at v_t.

Elevators

Elevators make great demonstration vehicles for exploring Newton's Laws of Motion, and are a popular topic on standardized physics exams. Typically, the demonstration involves a person standing on a scale in an elevator (for reasons unknown to most physics instructors).

In analyzing elevator problems, it is important first to realize that a scale does not directly read the weight of the person upon it. Rather, the scale provides a measure of the normal force the scale exerts. Once you realize that the reading on a scale is the normal force, analyzing problems involving elevators becomes a straightforward exercise in applying Newton's Laws of Motion.

Chapter 4: Dynamics

4.38 Q: A man with mass m stands on a scale in an elevator. If the scale reading is equal to mg when the elevator is at rest, what is the scale reading while the elevator is accelerating downwards with a magnitude of a?

(A) $m(g-a_y)$

(B) $m(g+a_y)$

(C) $m(a_y-g)$

(D) $m(a_y-g)$

4.38 A: Since the elevator is moving downward, set the positive y axis as pointing downward. We can then draw a free body diagram for the man in the elevator and apply Newton's 2nd Law in the y-direction.

$$\vec{F}_{net_y} = m\vec{a}_y \rightarrow m\vec{g} - \vec{F}_N = m\vec{a}_y \rightarrow$$

$$\vec{F}_N = m\vec{g} - m\vec{a}_y \rightarrow \vec{F}_N = m(\vec{g} - \vec{a}_y)$$

Therefore the correct answer is (A) $m(g-a_y)$.

4.39 Q: Lizzie stands on a scale in an elevator. If the scale on the elevator reads 600N when Lizzie is riding upward at a constant 4 m/s, what is the reading on the scale when the elevator is at rest?

(A) 420 N

(B) 600 N

(C) 780 N

(D) 840 N

4.39 A: (B) The scale reads the exact same at rest as it does while moving at constant velocity.

Ramps and Inclines

Now that you've developed an understanding of Newton's Laws of Motion, free body diagrams, friction, and forces on flat surfaces, you can extend these tools to situations on ramps, or inclined surfaces. The key to understanding these situations is creating an accurate free body diagram after choosing convenient x- and y-axes. Problem-solving steps are consistent with those developed for Newton's 2nd Law.

Take the example of a box on a ramp inclined at an angle of θ with respect to the horizontal. You can draw a basic free body diagram for this situation, with the force of gravity pulling the box straight down, the normal force perpendicular out of the ramp, and friction opposing motion (in this case pointing up the ramp).

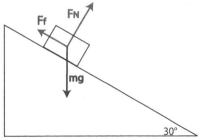

Once the forces acting on the box have been identified, you must be clever about our choice of x-axis and y-axis directions. Much like analyzing free falling objects and projectiles, if you set the positive x-axis in the direction of initial motion (or the direction the object wants to move if it is not currently moving), the y-axis must lie perpendicular to the ramp's surface (parallel to the normal force). Now, you can re-draw the free body diagram, this time superimposing it on the new axes. You may also want to rotate the book slightly counter-clockwise until the x- and y-axes are horizontal and vertical again.

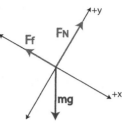

Unfortunately, the force of gravity on the box, *mg*, doesn't lie along one of the axes. Therefore, it must be broken up into components which do lie along the x- and y-axes in order to simplify the mathematical analysis. To do this, you can use geometry to break the weight down into a component parallel with the axis of motion (mg∥) and a component perpendicular to the x-axis (mg⊥) using trigonometry:

$$mg_{\parallel} = mg\sin(\theta)$$
$$mg_{\perp} = mg\cos(\theta)$$

You can now re-draw the free body diagram, replacing *mg* with its components. All the forces line up with the axes, making it straightforward to write Newton's 2nd Law Equations (F_{NETx} and F_{NETy}) and continue with your standard problem-solving strategy.

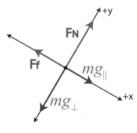

In the example shown with the modified free body diagram, you could write the Newton's 2nd Law Equations for both the x- and y-directions as follows:

$$F_{net_x} = mg_{\parallel} - F_f = mg\sin(\theta) - F_f = ma_x$$
$$F_{net_y} = F_N - mg_{\perp} = F_N - mg\cos(\theta) = 0$$

From this point, the problem becomes an exercise in algebra. If you need to tie the two equations together to eliminate a variable, don't forget the equation for the force of friction.

4.40 Q: Three forces act on a box on an inclined plane as shown in the diagram below. [Vectors are not drawn to scale.]

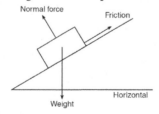

If the box is at rest, the net force acting on it is equal to

(A) the weight

(B) the normal force

(C) friction

(D) zero

4.40 A: (D) zero. If the box is at rest, the acceleration must be zero, therefore the net force must be zero.

4.41 Q: A 5-kg mass is held at rest on a frictionless 30° incline by force F. What is the magnitude of F?

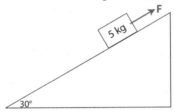

4.41 A: Start by identifying the forces on the box and making a free body diagram.

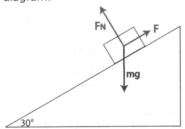

Break up the weight of the box into components parallel to and perpendicular to the ramp, and re-draw the free body diagram using the components of the box's weight.:

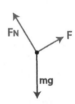

$$mg_{\parallel} = mg\sin(\theta)$$

$$mg_{\perp} = mg\cos(\theta)$$

Finally, use Newton's 2nd Law in the x-direction to solve for the force F.

$$F_{net} = F - mg_{\parallel} = F - mg\sin(\theta) = 0$$

$$F = mg\sin(\theta)$$

$$F = (5kg)(9.8\ ^{m}/_{s^2})\sin(30°) = 24.5N$$

4.42 Q: A 10-kg box slides down a frictionless 18° ramp. Find the acceleration of the box, and the time it takes the box to slide 2 meters down the ramp.

4.42 A: Start by identifying the forces on the box and making a free body diagram.

Break up the weight of the box into components parallel to and perpendicular to the ramp.

$$mg_{\parallel} = mg\sin(\theta)$$

$$mg_{\perp} = mg\cos(\theta)$$

Use Newton's 2nd Law to find the acceleration.

$$F_{net} = mg_{\parallel} = mg\sin(\theta) = ma$$

$$a = g\sin(\theta) = (9.8\,{}^m\!/_{s^2})(\sin(18°)) = 3.03\,{}^m\!/_{s^2}$$

Finally, use the acceleration to solve for the time it takes the box to travel 2m down the ramp using kinematic equations.

$$\Delta x = v_{0x}t + \tfrac{1}{2}a_x t^2 \rightarrow t = \sqrt{\frac{2\Delta x}{a}} = \sqrt{\frac{2(2m)}{3.03\,{}^m\!/_{s^2}}} = 1.15s$$

Note that the solution has no dependence on the mass of the box.

4.43 Q: The diagram below shows a 1.0 × 10⁵-newton truck at rest on a hill that makes an angle of 8.0° with the horizontal.

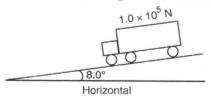

What is the component of the truck's weight parallel to the hill?

(A) 1.4 × 10³-newton

(B) 1.0 × 10⁴-newton

(C) 1.4 × 10⁴-newton

(D) 9.9 × 10⁴-newton

4.43 A: (C) 1.4 × 10⁴-newton

$$mg_{\parallel} = mg\sin(\theta) = (1.0 \times 10^5 \, N)(\sin(8°)) = 1.4 \times 10^4 \, N$$

4.44 Q: A block weighing 10 newtons is on a ramp inclined at 30° to the horizontal. A 3-newton force of friction, F_f , acts on the block as it is pulled up the ramp at constant velocity with force F, which is parallel to the ramp, as shown in the diagram below.

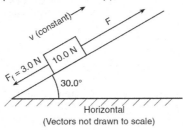

What is the magnitude of force F?

(A) 3 N

(B) 8 N

(C) 10 N

(D) 13 N

4.44 A: (B) Draw FBD and break weight of the box up into components.

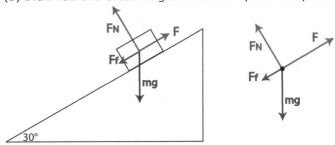

Break up the weight of the box into components parallel to and perpendicular to the ramp, then use Newton's second law, realizing that the forces must be balanced (net force is zero) since the box is moving at constant velocity.

$$F_{net} = F - F_f - mg\sin(\theta) = 0$$

$$F = F_f + mg\sin(\theta) = 3N + 10N \times \sin(30°) = 8N$$

4.45 Q: Which vector diagram best represents a cart slowing down as it travels to the right on a horizontal surface?

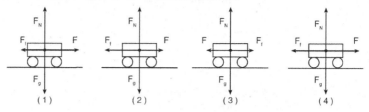

(1)   (2)   (3)   (4)

4.45 A: (2) Vertical forces are balanced, net force horizontally is in opposite direction of motion to create a negative acceleration and slow the cart down.

4.46 Q: The diagram represents a block at rest on an incline. Which diagram best represents the forces acting on the block? (F_f = frictional force, F_N = normal force, and F_w = weight.)

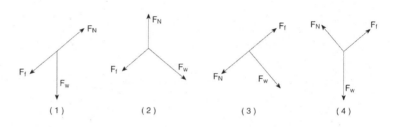

(1)   (2)   (3)   (4)

4.46 A: Correct Answer is (4).

Atwood Machines

An **Atwood Machine** is a basic physics laboratory device often used to demonstrate basic principles of dynamics and acceleration. The machine typically involves a pulley, a string, and a system of masses. Keys to solving Atwood Machine problems are recognizing that the force transmitted by a string or rope, known as tension, is constant throughout the string, and choosing a consistent direction as positive. Let's walk through an example to demonstrate.

4.47 Q: Two masses, m_1 and m_2, are hanging by a massless string from a frictionless pulley. If m_1 is greater than m_2, determine the acceleration of the two masses when released from rest.

4.47 A: First, identify a direction as positive. Since you can easily observe that m_1 will accelerate downward and m_2 will accelerate upward, since m_1 > m_2, call the direction of motion around the pulley and down toward m_1 the positive y direction. Then, you can create free body diagrams for both object m_1 and m_2, as shown below:

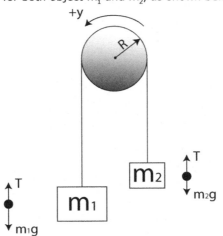

Using this diagram, write Newton's 2nd Law equations for both objects, taking care to note the positive y direction:

$$m_1g - T = m_1a \quad (1)$$
$$T - m_2g = m_2a \quad (2)$$

Next, combine the equations and eliminate T by solving for T in equation (B) and substituting in for T in equation (A).

$$T - m_2g = m_2a \qquad (2)$$
$$T = m_2g + m_2a \qquad (2b)$$
$$m_1g - m_2g - m_2a = m_1a \quad (1+2b)$$

Finally, solve for the acceleration of the system.

$$m_1g - m_2g - m_2a = m_1a \quad (1+2b)$$
$$m_1g - m_2g = m_1a + m_2a$$
$$g(m_1 - m_2) = a(m_1 + m_2)$$
$$a = g\frac{\left(m_1 - m_2\right)}{\left(m_1 + m_2\right)}$$

Alternately, you could treat both masses as part of the same system.

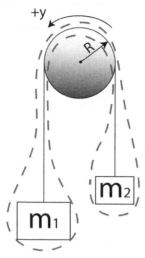

+y

R

m₂

m₁

Drawing a dashed line around the system, you can directly write an appropriate Newton's 2nd Law equation for the entire system.

$$m_1g - m_2g = (m_1 + m_2)a$$

$$a = g\frac{\left(m_1 - m_2\right)}{\left(m_1 + m_2\right)}$$

Note that if the string and pulley were not massless, this problem would get considerably more involved.

4.48 Q: Two masses are hung from a frictionless pulley by a massless spring. If m_1 is 5 kg, and m_2 is 7 kg, how far will m_2 fall in 2 seconds if released from rest?

4.48 A: This is both a dynamics and a kinematics problem. To solve this problem, you must first find the acceleration of m_2. Once you know the acceleration, you can use kinematics to determine how far m_2 falls in the given time interval.

In this problem, it is obvious that m_2 will accelerate toward the ground while m_1 accelerates upward, so choose the positive y-direction accordingly.

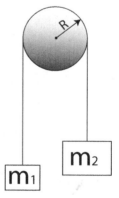

R

m₂

m₁

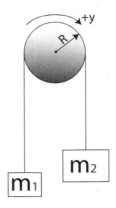

Looking at this from the systems approach, you can write the Newton's 2nd Law Equation as follows:

$$F_{NET_y} = m_2 g - m_1 g = (m_1 + m_2)a$$

$$a = g \frac{(m_2 - m_1)}{(m_2 + m_1)}$$

$$a = (9.8\,{}^m\!/_{s^2}) \times \frac{(7kg - 5kg)}{(7kg + 5kg)}$$

$$a = 1.63\,{}^m\!/_{s^2}$$

Now, knowing the acceleration of the system, find how far m_2 falls in two seconds using the kinematic equations.

VERTICAL MOTION TABLE

Variable	Value
v_0	0 m/s
v	?
Δy	FIND
a	1.63 m/s²
t	2 s

With the given and find information clearly defined, you can utilize the kinematic equations to determine how far the mass fell in two seconds.

$$\Delta y = v_0 t + \tfrac{1}{2}at^2$$

$$\Delta y = 0 + \tfrac{1}{2}(1.63\,{}^m\!/_{s^2})(2s)^2$$

$$\Delta y = 3.27m$$

4.49 Q: Two masses, m_1 and m_2, are connected by a light string over a massless pulley as shown. Assuming a frictionless surface, find the acceleration of m_2.

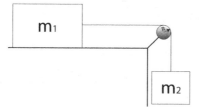

4.49 A: First, draw a free body diagram for each of the two masses. Call the clockwise direction around the pulley the positive y direction.

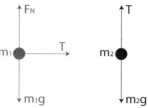

Next, write Newton's 2nd Law equations for each mass and solve for the acceleration of m_2, recognizing that the magnitude of the horizontal acceleration of m_1 must equal the magnitude of the vertical acceleration of m_2. Because the pulley is massless, you can state that the tension must be uniform everywhere in the light string.

$$F_{net_X} = T = m_1 a$$

$$F_{net_Y} = m_2 g - T = m_2 a \xrightarrow{T = m_1 a}$$

$$m_2 g - m_1 a = m_2 a \rightarrow a = g\left(\frac{m_2}{m_1 + m_2}\right)$$

Test Your Understanding

1. Can an object exert a force upon itself? Explain with an example.

2. Design at least two separate experiments to measure the acceleration due to gravity on Earth's surface using Newton's 2nd Law of Motion.

3. For the following situations, identify the appropriate force pairs: batter hitting a baseball; moon orbiting Earth; tree falling toward the ground; a skydiver falling at terminal velocity.

4. An astronaut and a satellite are floating through the outer reaches of space, where gravitational forces from outside objects are negligible. Assuming the astronaut and a satellite comprise a system, is it possible to change the center-of-mass velocity of the system? Explain.

5. Design an experiment to determine both the kinetic and static coefficients of friction between two surfaces. Try it out for a number of surfaces. Do any of the results surprise you? Explain.

Chapter 4: Dynamics

Chapter 5: Work, Power & Energy

*"Ambition is like a vector; it needs magnitude and direction.
Otherwise, it's just energy."*

— *Grace Lindsay*

Objectives

1. Define work and calculate the work done by a force.
2. Apply relationships between work, net force, displacement, velocity and kinetic energy to solve a variety of problems.
3. Calculate the power of a system.
4. Identify, describe, and calculate the potential energy of a system.
5. Apply conservation of energy to analyze energy transitions and transformations in a system.
6. Analyze the relationship between work done on or by a system, and the energy gained or lost by that system.
7. Use Hooke's Law to determine the elastic force on an object.
8. Calculate a system's elastic potential energy.

Work, energy and power are highly inter-related concepts that come up regularly in everyday life. You do work on an object when you move it. The rate at which you do the work is your power output. When you do work on an object, you transfer energy from one object to another. In this chapter you'll explore how energy is transferred and transformed, how doing work on an object changes its energy, and how quickly work can be done.

Work

Sometimes you work hard. Sometimes you're a slacker. But, right now, are you doing work? And what is meant by the word "work?" In physics terms, **work** is the process of moving an object by applying a force, or, more formally, work is the energy transferred by an external force exerted on an object or system that moves the object or system.

I'm sure you can think up countless examples of work being done, but a few that spring to mind include pushing a snowblower to clear the driveway, pulling a sled up a hill with a rope, stacking boxes of books from the floor onto a shelf, and throwing a baseball from the pitcher's mound to home plate.

Let's take a look at a few scenarios and investigate what work is being done.

In the first scenario, a monkey in a jet pack blasts through the atmosphere, accelerating to higher and higher speeds. In this case, the jet pack is applying a force causing it to move. But what is doing the work? Hot expanding gases are pushed backward out of the jet pack. Using Newton's 3rd Law, you observe the reactionary force of the gas pushing the jet pack forward, causing a displacement. Therefore, the expanding exhaust gas is doing work on the jet pack.

In the second scenario, a girl struggles to push her stalled car, but can't make it move. Even though she's expending significant effort, no work is being done on the car because it isn't moving.

In the final scenario, a child in a ghost costume carries a bag of Halloween candy across the yard. In this situation, the child applies a force upward on the bag, but the bag moves horizontally. From this perspective, the forces of the child's arms on the bag don't cause the displacement, therefore no work is being done by the child.

It's important to note that when calculating work, only the force applied in the direction of the object's displacement counts! This means that if the force and displacement vectors aren't in exactly the same direction, you need to take the component of force in the direction of the object's displacement.

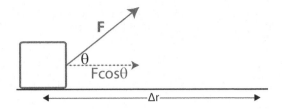

To do this, line up the force and displacement vectors tail-to-tail and measure the angle between them. Since this component of force can be calculated by multiplying the force by the cosine of the angle between the force and displacement vectors, you can write the work equation as:

$$W = \Delta E = F_{\|}r = Fr\cos\theta$$

W is the work done, **F** is the force applied in newtons, **r** is the object's displacement in meters, and theta is the angle between **F** and **r**. The equation as a whole states that the work done is equal to the energy transferred, which is equal to the component of force parallel to the displacement multiplied by the displacement, which is equal to the magnitude of the force, multiplied by the magnitude of the displacement, multiplied by the cosine of the angle between them. When the work done on an object is independent of the object's path, the force doing the work is known as a **conservative force**.

The units of work can be found by performing unit analysis on the work formula. If work is force multiplied by distance, the units must be the units of force multiplied by the units of distance, or newtons multiplied by meters. A newton-meter is also known as a Joule (J).

5.01 Q: An appliance salesman pushes a refrigerator 2 meters across the floor by applying a force of 200N. Find the work done.

5.01 A: Since the force and displacement are in the same direction, the angle between them is 0.
$$W = F_{\|}r = Fr\cos\theta = (200N)(2m)\cos 0 = 400J$$

5.02 Q: A friend's car is stuck on the ice. You push down on the car to provide more friction for the tires (by way of increasing the normal force), allowing the car's tires to propel it forward 5m onto less slippery ground. How much work did you do?

5.02 A: You applied a downward force, yet the car's displacement was sideways. Therefore, the angle between the force and displacement vectors is 90°.
$$W = F_{\|}r = Fr\cos\theta = Fr\cos 90 = 0$$

5.03 Q: You push a crate up a ramp with a force of 10N. Despite your pushing, however, the crate slides down the ramp a distance of 4m. How much work did you do?

5.03 A: Since the direction of the force you applied is opposite the direction of the crate's displacement, the angle between the two vectors is 180°.

$$W = F_{\parallel}r = Fr\cos\theta = (10N)(4m)\cos 180 = -40J$$

5.04 Q: How much work is done in lifting an 8-kg box from the floor to a height of 2m above the floor?

5.04 A: It's easy to see the displacement is 2m, and the force must be applied in the direction of the displacement, but what is the force? To lift the box you must match and overcome the force of gravity on the box. Therefore, the force applied is equal to the gravitational force, or weight, of the box, mg=(8kg)(9.8m/s²)=78.4N.

$$W = F_{\parallel}r = Fr\cos\theta = (78.4N)(2m)\cos 0 = 157J$$

5.05 Q: Barry and Sidney pull a 30-kg wagon with a force of 500N a distance of 20m. The force acts at a 30° angle to the horizontal. Calculate the work done.

5.05 A: $W = F_{\parallel}r = Fr\cos\theta = (500N)(20m)\cos 30° = 8660J$

5.06 Q: The work done in lifting an apple one meter near Earth's surface is approximately

(A) 1 J

(B) 0.01 J

(C) 100 J

(D) 1000 J

5.06 A: (A) The trick in this problem is recalling the approximate weight of an apple. With an "order-of-magnitude" estimate, you can say an apple has a mass of 0.1 kg, or a weight of 1 N. Given this information, the work done is:

$$W = F_{\parallel}r = Fr\cos\theta = (1N)(1m)\cos 0° = 1J$$

5.07 Q: As shown in the diagram, a child applies a constant 20-newton force along the handle of a wagon which makes a 25° angle with the horizontal.

How much work does the child do in moving the wagon a horizontal distance of 4.0 meters?

(A) 5.0 J

(B) 34 J

(C) 73 J

(D) 80. J

5.07 A: (D) $W = F_\parallel r = Fr\cos\theta = (20N)(4m)\cos 25° = 73J$

The area under a force vs. displacement graph is the work done by the force. Consider the situation of a block being pulled across a table with a constant force of 5 Newtons over a displacement of 5 meters, then the force gradually tapers off over the next 5 meters.

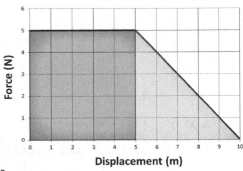

The work done by the force moving the block can be calculated by taking the area under the force vs. displacement graph (a combination of a rectangle and triangle) as follows:

$$Work = Area_{rectangle} + Area_{triangle}$$
$$Work = lw + \frac{1}{2}bh$$
$$Work = (5m)(5N) + \frac{1}{2}(5m)(5N)$$
$$Work = 37.5J$$

5.08 Q: A boy pushes his wagon at constant speed along a level sidewalk. The graph below represents the relationship between the horizontal force exerted by the boy and the distance the wagon moves.

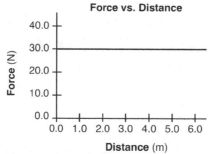

Force vs. Distance

What is the total work done by the boy in pushing the wagon 4.0 meters?

(A) 5.0 J

(B) 7.5 J

(C) 120 J

(D) 180 J

5.08 A: (C) 120 J $\quad Work = Area_{rectangle} = lw = (4m)(30N) = 120J$

5.09 Q: A box is wheeled to the right with a varying horizontal force. The graph below represents the relationship between the applied force and the distance the box moves.

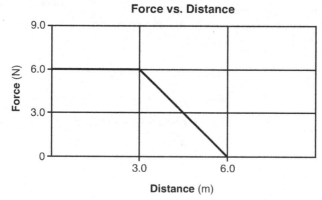

Force vs. Distance

What is the total work done in moving the box 6 meters?

(A) 9.0 J

(B) 18 J

(C) 27 J

(D) 36 J

5.09 A: (C) $Work = Area_{rectangle} + Area_{triangle}$

$Work = lw + \frac{1}{2}bh$

$Work = (3m)(6N) + \frac{1}{2}(3m)(6N)$

$Work = 27\,J$

5.10 Q: An 80-kg wooden box is pulled 10 meters horizontally across a wood floor at a constant velocity by a 250-newton force at an angle of 37 degrees above the horizontal. If the coefficient of kinetic friction between the floor and the box is 0.315, find the work done by friction.

5.10 A: First, draw a diagram of the situation.

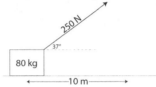

Next, create a FBD and pseudo-FBD detailing the forces on the box.

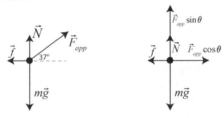

Then, write a Newton's 2nd Law equation in the x-direction and solve for the force of friction, recognizing that the acceleration of the box must be zero since it moves at constant velocity.

$$F_{net_x} = F_{app}\cos\theta - f = ma_x \xrightarrow{a=0} 250N\cos(37°) - f = 0 \rightarrow$$

$$f = 200N$$

Finally, solve for the work done by friction.

$$W_f = F_f r\cos\theta = (200N)(10m)\cos(180°) = -2000J$$

5.11 Q: Four carts, initially at rest on a flat surface, are subjected to varying forces as the carts move to the right a set distance, depicted
in the diagram below.

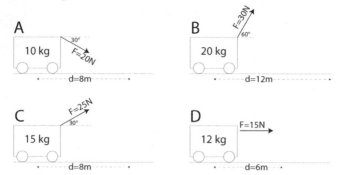

Rank the four carts from least to greatest in terms of

I) work done by the applied force on the carts

II) inertia

III) normal force applied by the surface to the carts

5.11 A: I) D, A, C, B

II) A, D, C, B

III) A, D, C, B

Power

Power is a term used quite regularly in all aspects of life. People talk about
how powerful the new boat motor is, the power of positive thinking, and even
the power company's latest bill. All of these uses
of the term power relate to how much work can
be done in some amount of time, or the
rate at which energy is transferred.

In physics, work can be defined in two ways. Work is the process of moving an object by applying a force. The rate at which the force does work is
known as **power** (P). Work is also the transfer of energy, so power is also
the rate at which energy is transferred into, out of, or within a system. The
units of power are the units of work divided by time, or Joules per second,
known as a **Watt** (W).

$$P = \frac{W}{t} = \frac{\Delta E}{t}$$

Since power is the rate at which work is done, it is possible to have the same
amount of work done but with a different supplied power, if the time is different.

5.12 Q: Rob and Peter move a sofa 3 meters across the floor by applying a combined force of 200N horizontally. If it takes them 6 seconds to move the sofa, what amount of power did they supply?

5.12 A: $P = \dfrac{W}{t} = \dfrac{Fr\cos\theta}{t} = \dfrac{(200N)(3m)}{6s} = 100W$

5.13 Q: Kevin then pushes the same sofa 3 meters across the floor by applying a force of 200N. Kevin, however, takes 12 seconds to push the sofa. What amount of power did Kevin supply?

5.13 A: $P = \dfrac{W}{t} = \dfrac{Fr\cos\theta}{t} = \dfrac{(200N)(3m)}{12s} = 50W$

As you can see, although Kevin did the same amount of work as Rob and Peter in pushing the sofa (600J), Rob and Peter supplied twice the power of Kevin because they did the same work in half the time!

There's more to the story, however. Since power is defined as work over time, and because work is equal to force (in the direction parallel to the displacement) multiplied by displacement, you can replace work in the equation with F×r×cos(θ):

$$P = \frac{W}{t} = \frac{Fr\cos\theta}{t}$$

Looking carefully at this equation, you can observe a displacement divided by time. Since displacement divided by time is the definition of average velocity, you can replace Δr/t with v in the equation and, assuming the force is in the direction of the displacement (cosθ=1) you obtain:

$$P = \frac{W}{t} = \frac{Fr\cos\theta}{t} = F\bar{v}$$

So, not only is power equal to work done divided by the time required, it's also equal to the force applied (in the direction of the displacement) multiplied by the average velocity of the object.

5.14 Q: Motor A lifts a 5000N steel crossbar upward at a constant 2 m/s. Motor B lifts a 4000N steel support upward at a constant 3 m/s. Which motor is supplying more power?

5.14 A: Motor B supplies more power than Motor A.

$$P_{MotorA} = F\overline{v} = (5000N)(2\,{}^m\!/_s) = 10000W$$
$$P_{MotorB} = F\overline{v} = (4000N)(3\,{}^m\!/_s) = 12000W$$

5.15 Q: A 70-kilogram cyclist develops 210 watts of power while pedaling at a constant velocity of 7 meters per second east. What average force is exerted eastward on the bicycle to maintain this constant speed?

(A) 490 N

(B) 30 N

(C) 3.0 N

(D) 0 N

5.15 A: (B) $P = F\overline{v}$

$$F = \frac{P}{\overline{v}} = \frac{210W}{7\,{}^m\!/_s} = 30N$$

5.16 Q: Alien A lifts a 500-newton child from the floor to a height of 0.40 meters in 2 seconds. Alien B lifts a 400-newton student from the floor to a height of 0.50 meters in 1 second. Compared to Alien A, Alien B does

(A) the same work but develops more power

(B) the same work but develops less power

(C) more work but develops less power

(D) less work but develops more power

5.16 A: (A) the same work but develops more power.

5.17 Q: A 110-kilogram bodybuilder and his 55-kilogram friend run up identical flights of stairs. The bodybuilder reaches the top in 4.0 seconds while his friend takes 2.0 seconds. Compared to the power developed by the bodybuilder while running up the stairs, the power developed by his friend is

(A) the same

(B) twice as much

(C) half as much

(D) four times as much

5.17 A: (A) the same.

5.18 Q: Mary holds a 5-kg mirror against the wall 1.5 meters above the ground for 20 seconds while Bob nails it in place. What is Mary's power output during that time period?

(A) 2.45 Watts

(B) 3.68 Watts

(C) 66.7 Watts

(D) None of the above

5.18 A: (D) There is no power output because the work done is zero. There is no displacement of the mirror.

5.19 Q: Which of the following are appropriate units for power? Choose all that apply.

(A) $\dfrac{J}{s}$

(C) $\dfrac{kg \bullet m^2}{s^2}$

(B) $\dfrac{N \bullet m^2}{s}$

(D) $\dfrac{kg \bullet m^2}{s^3}$

5.19 A: (A) and (D) are appropriate units for power.

5.20 Q: A box of mass m is pushed up a ramp at constant velocity v to a maximum height h in time t by force F. The ramp makes an angle of θ with the horizontal as shown in the diagram below.

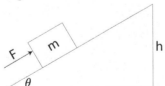

What is the power supplied by the force? Choose all that apply.

(A) $\dfrac{mgh}{t}$

(C) $\dfrac{Fh}{t\sin\theta}$

(B) $\dfrac{mgh}{t\sin\theta}$

(D) Fv

5.20 A: (C) and (D) are both correct expressions for the power supplied by the force.

Energy

We've all had days where we've had varying amounts of energy. You've gotten up in the morning, had to drag yourself out of bed, force yourself to get ready for school, and once you finally get to class, you don't have the energy to do much work. Other days, when you've had more energy, you may have woken up before the alarm clock, hustled to get ready for the day while a bunch of thoughts bounced around in your head, and hurried on to begin your activities. Then, throughout the day, the more work you do, the more energy you lose... What's the difference in these days?

In physics, **energy** is the ability or capacity to do work. And as mentioned previously, work is the process of moving an object. So, if you combine the definitions, energy is the ability or capacity to move an object. So far you've examined kinetic energy, or energy of motion, and therefore kinetic energy must be the ability or capacity of a moving object to move another object! Mathematically, kinetic energy is calculated using the formula:

$$K = \tfrac{1}{2}mv^2$$

Of course, there are more types of energy than just kinetic. Energy comes in many forms, which you can classify as kinetic (energy of motion) or potential (stored) to various degrees. This includes solar energy, thermal energy, gravitational potential energy, nuclear energy, chemical potential energy, sound energy, electrical energy, elastic potential energy, light energy, and so on. In all cases, energy can be transformed from one type to another and you can transfer energy from one object to another by doing work.

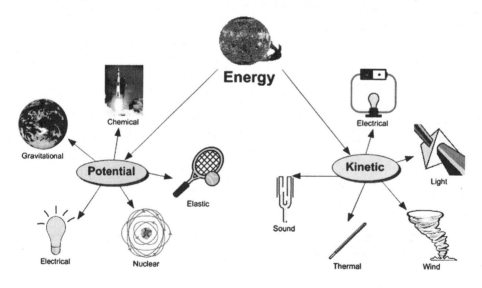

The units of energy are the same as the units of work, joules (J). Through dimensional analysis, observe that the units of KE (kg·m²/s²) must be equal to the units of work (N·m):

$$\frac{kg \bullet m^2}{s^2} = N \bullet m = J$$

Gravitational Potential Energy

Potential energy is energy an object possesses due to its position or condition. In order for an object to have potential energy, it must interact with another object or system. An object in isolation can have only kinetic energy.

Gravitational potential energy is the energy an object possesses because of its position in a gravitational field (height), where another object or system is providing the gravitational field.

Assume a 10-kilogram box sits on the floor. You can arbitrarily call its current potential energy zero, just to give a reference point. If you do work to lift the box one meter off the floor, you need to overcome the force of gravity on the box (its weight) over a distance of one meter. Therefore, the work you do on the box can be obtained from:

$$W = F_{\parallel}r = (mg)\Delta y = (10kg)(9.8\,{}^{m}\!/_{s^2})(1m) = 98J$$

So, to raise the box to a height of 1m, you must do 98 Joules of work on the box. The work done in lifting the box is equal to the change in the potential energy of the box, so the box's gravitational potential energy must increase by 98 Joules.

When you performed work on the box, you transferred some of your stored energy to the box. Along the way, it just so happens that you derived the formula for the change in gravitational potential energy of an object. In a constant gravitational field, the change in the object's gravitational potential energy, ΔU_g, is equal to the force of gravity on the box (mg) multiplied by its change in height, Δy.

$$\Delta U_g = mg\Delta y$$

This formula can be used to solve a variety of problems involving the potential energy of an object.

5.21 Q: The diagram below represents a 155-newton box on a ramp. Applied force F causes the box to slide from point A to point B.

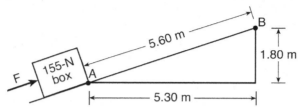

What is the total amount of gravitational potential energy gained by the box?

(A) 28.4 J

(B) 279 J

(C) 868 J

(D) 2740 J

5.21 A: (B) $\Delta U_g = mg\Delta y = (155N)(1.8m) = 279J$

5.22 Q: Which situation describes a system with decreasing gravitational potential energy?

(A) a girl stretching a horizontal spring

(B) a bicyclist riding up a steep hill

(C) a rocket rising vertically from Earth

(D) a boy jumping down from a tree limb

5.22 A: (D) The boy's height above ground is decreasing, so his gravitational potential energy is decreasing.

5.23 Q: Which is an SI unit for energy?

(A) $\dfrac{kg \bullet m^2}{s^2}$ (C) $\dfrac{kg \bullet m}{s}$

(B) $\dfrac{kg \bullet m^2}{s}$ (D) $\dfrac{kg \bullet m}{s^2}$

5.23 A: (A) is equivalent to a newton-meter, also known as a Joule.

5.24 Q: A car travels at constant speed v up a hill from point A to point B, as shown in the diagram below.

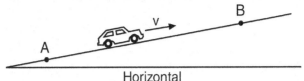

As the car travels from A to B, its gravitational potential energy

(A) increases and its kinetic energy decreases

(B) increases and its kinetic energy remains the same

(C) remains the same and its kinetic energy decreases

(D) remains the same and its kinetic energy remains the same

5.24 A: (B) The car's height above ground increases so gravitational potential energy increases, and velocity remains constant, so kinetic energy remains the same. Note that the car's engine must do work to maintain a constant velocity.

5.25 Q: An object is thrown vertically upward. Which pair of graphs best represents the object's kinetic energy and gravitational potential energy as functions of its displacement while it rises?

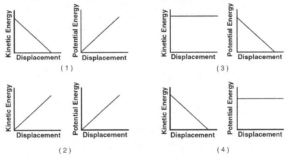

5.25 A: (1) shows the object's kinetic energy decreasing as it slows down on its way upward, while its potential energy increases as its height increases.

5.26 Q: While riding a chairlift, a 55-kilogram snowboarder is raised a vertical distance of 370 meters. What is the total change in the snowboarder's gravitational potential energy?

(A) 5.4×10^1 J

(B) 5.4×10^2 J

(C) 2.0×10^4 J

(D) 2.0×10^5 J

5.26 A: (D) 2.0×10^5 J

$$\Delta U_g = mg\Delta y = (55kg)(9.8\,{}^m\!/_{s^2})(370m) = 2 \times 10^5\,J$$

5.27 Q: A pendulum of mass M swings on a light string of length
L as shown in the diagram at right. If the mass hang-
ing directly down is set as the zero point of gravitational potential energy, find the gravitational potential
energy of the pendulum as a function of θ and L.

5.27 A: First use some basic geometry and trigonometry to
analyze the situation and set up the problem. Re-
draw the diagram as shown at right, then, solve for
the change in vertical displacement, Δy.

$$\cos\theta = \frac{adj}{hyp} \rightarrow adj = hyp \bullet \cos\theta = L\cos\theta$$

$$\Delta y = L - adj = L - L\cos\theta = L(1-\cos\theta)$$

Now, utilize your value for Δy to find the gravitational potential
energy above your reference point.

$$\Delta U_g = mg\Delta y \xrightarrow{\Delta y = L(1-\cos\theta)} mgL(1-\cos\theta)$$

Springs and Hooke's Law

An interesting application of work combined with the Force and Displacement graph is examining the force applied by a spring. The more
you stretch a spring, the greater the force of the spring. Similarly,
the more you compress a spring, the greater the force. This can be
modeled as a linear relationship known as Hooke's Law, where the
magnitude of the force applied by the spring is equal to a constant
multiplied by the magnitude of the displacement of the spring.

$$|\vec{F_s}| = k|\vec{x}|$$

F_s is the force of the spring in newtons, x is the displacement of the spring
from its equilibrium (or rest) position, in meters, and k is the spring constant,
which tells you how stiff or powerful a spring is, in newtons per meter. The
larger the spring constant, k, the more force the spring applies per amount
of displacement. In some cases you may see this written as F_s=-kx, where
the negative sign indicates the force is in the opposite direction of the displacement.

You can determine the spring constant of a spring by making a graph of the force from a spring on the y-axis, and placing the displacement of the spring from its equilibrium, or rest position, on the x-axis. The slope of the graph will give you the spring constant. For the case of the spring depicted in the graph at right, you can find the spring constant as follows:

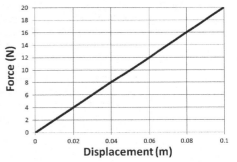

$$k = Slope = \frac{rise}{run} = \frac{\Delta F}{\Delta x} = \frac{20N - 0N}{0.1m - 0m} = 200 \, ^N/_m$$

You must have done work to compress or stretch the spring, since you applied a force and caused a displacement. You can find the work done in stretching or compressing a spring by taking the area under the graph. For the spring shown, to displace the spring 0.1m, you can find the work done as shown below:

$$Work = Area_{tri} = \tfrac{1}{2}bh = \tfrac{1}{2}(0.1m)(20N) = 1J$$

5.28 Q: In an experiment, a student applied various forces to a spring and measured the spring's corresponding elongation. The table at right shows his data.

Plot force versus elongation and draw the best-fit line. Then, using your graph, calculate the spring constant of the spring. Show all your work.

Force (newtons)	Elongation (meters)
0	0
1.0	0.30
3.0	0.67
4.0	1.00
5.0	1.30
6.0	1.50

5.28 A:

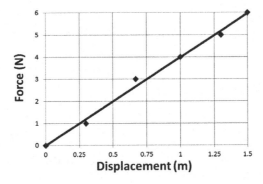

$$k = Slope = \frac{rise}{run} = \frac{\Delta F}{\Delta x} = \frac{6N - 0N}{1.5m - 0m} = 4 \, ^N/_m$$

5.29 Q: In a laboratory investigation, a student applied various downward forces to a vertical spring. The applied forces and the corresponding elongations of the spring from its equilibrium position are recorded in the data table.

Construct a graph, marking an appropriate scale on the axis labeled "Force (N)." Plot the data points for force versus elongation. Draw the best-fit line or curve. Then, using your graph, calculate the spring constant of this spring. Show all your work.

Force (newtons)	Elongation (meters)
0	0
0.5	0.010
1.0	0.018
1.5	0.027
2.0	0.035
2.5	0.046

5.29 A:

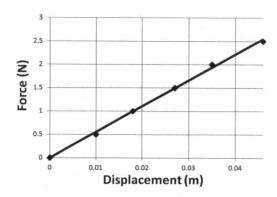

$$k = Slope = \frac{rise}{run} = \frac{\Delta F}{\Delta x} = \frac{2.5N - 0.8N}{0.046m - 0.015m} = 55\,{}^{N}\!/_{m}$$

5.30 Q: A 10-newton force compresses a spring 0.25 meter from its equilibrium position. Calculate the spring constant of this spring.

5.30 A: $\left|\vec{F}_{S}\right| = k\left|\vec{x}\right| \rightarrow k = \frac{\left|\vec{F}_{S}\right|}{\left|\vec{x}\right|} = \frac{10N}{0.25m} = 40\,{}^{N}\!/_{m}$

Another form of potential energy involves the stored energy an object possesses due to its position in a stressed elastic system. An object at the end of a compressed spring, for example, has **elastic potential energy**. When the spring is released, the elastic potential energy of the spring will do work on the object, moving the object and transferring the energy of the spring into kinetic energy of the object. Other examples of elastic potential energy include tennis rackets, rubber bands, bows (as in bows and arrows), trampolines, bouncy balls, and even pole-vaulting poles.

The most common problems involving elastic potential energy in introductory physics involve the energy stored in a spring. As you learned previously, the force needed to compress or stretch a spring from its equilibrium position increases linearly. The more you stretch or compress the spring, the more force it applies trying to restore itself to its equilibrium position. This can be modeled using Hooke's Law:

$$\left|\vec{F}_s\right| = k\left|\vec{x}\right|$$

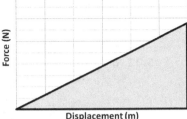

Force (N)

Displacement (m)

Further, you can find the work done in compressing or stretching the spring by taking the area under a force vs. displacement graph for the spring.

$$W = F_{\parallel}r = Area_{triangle} = \tfrac{1}{2}bh = \tfrac{1}{2}(x)(kx) = \tfrac{1}{2}kx^2$$

Since the work done in compressing or stretching the spring from its equilibrium position transfers energy to the spring, you can conclude that the potential energy stored in the spring must be equal to the work done to compress the spring. The potential energy of a spring (U_S) is therefore given by:

$$U_S = \tfrac{1}{2}kx^2$$

5.31 Q: A spring with a spring constant of 4.0 newtons per meter is compressed by a force of 1.2 newtons. What is the total elastic potential energy stored in this compressed spring?

(A) 0.18 J

(B) 0.36 J

(C) 0.60 J

(D) 4.8 J

5.31 A: (A) Us can't be calculated directly since x isn't known, but x can be found from Hooke's Law:

$$\left|\vec{F}_s\right| = k\left|\vec{x}\right| \rightarrow \left|\vec{x}\right| = \frac{\left|\vec{F}_s\right|}{k} = \frac{1.2N}{4\,{}^N\!/_m} = 0.3m$$

With x known, the potential energy equation for a spring can be utilized.

$$U_S = \tfrac{1}{2}kx^2 = \tfrac{1}{2}(4\,{}^N\!/_m)(0.3m)^2 = 0.18J$$

5.32 Q: An unstretched spring has a length of 10 centimeters. When the spring is stretched by a force of 16 newtons, its length is increased to 18 centimeters. What is the spring constant of this spring?

(A) 0.89 N/cm

(B) 2.0 N/cm

(C) 1.6 N/cm

(D) 1.8 N/cm

5.32 A: (B) $\left|\vec{F}_s\right| = k\left|\vec{x}\right| \rightarrow k = \dfrac{\left|\vec{F}_s\right|}{\left|\vec{x}\right|} = \dfrac{16N}{8cm} = 2.0\,{}^{N}\!/_{cm}$

5.33 Q: Which graph best represents the relationship between the elastic potential energy stored in a spring and its elongation from equilibrium?

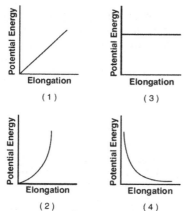

5.33 A: (2) due to the displacement² relationship.

5.34 Q: A pop-up toy has a mass of 0.020 kilogram and a spring constant of 150 newtons per meter. A force is applied to the toy to compress the spring 0.050 meter.

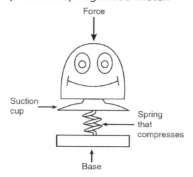

(A) Calculate the potential energy stored in the compressed spring.

(B) The toy is activated and all the compressed spring's potential energy is converted to gravitational potential energy. Calculate the maximum vertical height to which the toy is propelled.

5.34 A: (A) $U_s = \frac{1}{2}kx^2 = \frac{1}{2}(150 \, {}^{N}\!/_{m})(0.05m)^2 = 0.1875J$

(B) $U_g = mg\Delta y \rightarrow \Delta y = \dfrac{U_g}{mg} = \dfrac{0.1875J}{(0.02kg)(9.8 \, {}^{m}\!/_{s^2})} = 0.96m$

5.35 Q: A spring with a spring constant of 80 newtons per meter is displaced 0.30 meter from its equilibrium position. The potential energy stored in the spring is

(A) 3.6 J

(B) 7.2 J

(C) 12 J

(D) 24 J

5.35 A: (A) $U_s = \frac{1}{2}kx^2 = \frac{1}{2}(80 \, {}^{N}\!/_{m})(0.3m)^2 = 3.6J$

Work-Energy Theorem

Of course, there are many different kinds of energy which haven't been mentioned specifically. Energy can be converted among its many different forms, such as mechanical (which is kinetic, gravitational potential, and elastic potential), electromagnetic, nuclear, and thermal (or internal) energy.

When a force does work on a system, the work done changes the system's energy. If the work done increases motion, there is an increase in the system's kinetic energy. If the work done increases the object's height, there is an increase in the system's gravitational potential energy. If the work done compresses a spring, there is an increase in the system's elastic potential energy. If the work is done against friction, however, where does the energy go? In this case, the energy isn't lost, but instead increases the rate at which molecules in the object vibrate, increasing the object's temperature, or internal energy.

The internal energy of a system includes the kinetic energy of the objects that make up the system and the potential energy of the configuration of the objects that make up the system.

The understanding that the work done on a system by an external force changes the energy of the system is known as the Work-Energy Relationship. If an external force does positive work on the system, the system's total energy increases. If, instead, the system does work, the system's total energy decreases. Put another way, you add energy to a system by doing work on it and take energy from a system when the system does the work (much like you add value to your bank account by making a deposit and take value from your account by writing a check).

When the force applied on an object is parallel to the object's displacement, the work done increases the kinetic energy of the object. When the force applied is opposite the direction of the object's displacement, the work done decreases the kinetic energy of the object. This is known as the **Work-Energy Theorem**.

5.36 Q: Given the following sets of velocity and net force vectors for a given object, state whether you expect the kinetic energy of the object to increase, decrease, or remain the same.

$$\vec{v} \quad \vec{F}_{net}$$

(A) →   ←

(B) →   ↑

(C) ↗   ↘

(D) ↖   →

5.36 A: (A) decrease

(B) remain the same (no work done as long as v and F are perpendicular)

(C) increase

(D) decrease

5.37 Q: A chef pushes a 10-kilogram pastry cart from rest a distance of 5 meters with a constant horizontal force of 10 N. Assuming a frictionless surface, determine the cart's change in kinetic energy and its final velocity.

5.37 A: First find the work done by the chef, which will be equal to the cart's change in kinetic energy.

$$W = F_{\parallel}r = (10N)(5m) = 50J$$

Next, solve for the cart's final velocity.

$$K = \tfrac{1}{2}mv^2 \rightarrow v = \sqrt{\frac{2K}{m}} = \sqrt{\frac{2(50J)}{10kg}} = 3.2\,{m}\!/\!{s}$$

5.38 Q: A pitcher throws a 143-gram baseball toward the catcher at 45 m/s. If the catcher's hand moves back a distance of 6 cm in stopping the ball, determine the average force exerted on the catcher's hand.

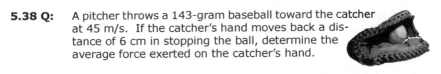

5.38 A: There are several paths to solving this problem, including application of the Work-Energy Theorem.

$$W = \Delta K \rightarrow F \bullet r = \tfrac{1}{2}mv^2 \rightarrow F = \frac{mv^2}{2r} = \frac{(0.143kg)(45^m/_s)^2}{2(0.06m)} = 2415N$$

5.39 Q: In the following diagrams, a force F acts on a cart in motion on a frictionless surface to change its velocity. The initial velocity of the cart and final velocity of each cart are shown. You do not know how far or in which direction the cart traveled. Rank the energy required to change each cart's velocity from greatest to least.

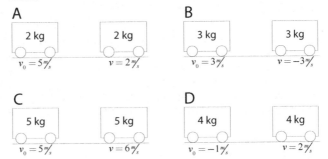

A

| 2 kg | 2 kg |
| $v_0 = 5^m/_s$ | $v = 2^m/_s$ |

B

| 3 kg | 3 kg |
| $v_0 = 3^m/_s$ | $v = -3^m/_s$ |

C

| 5 kg | 5 kg |
| $v_0 = 5^m/_s$ | $v = 6^m/_s$ |

D

| 4 kg | 4 kg |
| $v_0 = -1^m/_s$ | $v = 2^m/_s$ |

5.39 A: C, B, A, D (First find the changes in K, which must equal the work done according to the Work-Energy Theorem. Be careful of passing through v=0!)

5.40 Q: Given the force vs. displacement graph below for a net force applied horizontally to an object of mass m initially at rest on a frictionless surface, determine the object's final speed in terms of F_{max}, r_1, r_2, r_3, and m. You may assume the force does not change its direction.

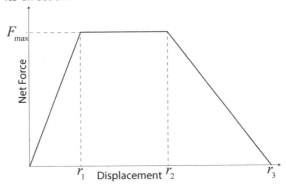

5.40 A: The work done is the area under the graph, which, according to the Work-Energy Theorem, is also equal to the change in the object's kinetic energy, therefore you can set the area of the graph equal to the object's final kinetic energy given the initial kinetic energy is zero.

$$W = Area = \tfrac{1}{2}bh + lw + \tfrac{1}{2}bh = \tfrac{1}{2}mv^2 \rightarrow$$

$$\tfrac{1}{2}r_1 F_{max} + (r_2 - r_1)F_{max} + \tfrac{1}{2}(r_3 - r_2)F_{max} = \tfrac{1}{2}mv^2 \rightarrow$$

$$v^2 = \frac{F_{max}}{m}(r_3 + r_2 - r_1) \rightarrow v = \sqrt{\frac{F_{max}}{m}(r_3 + r_2 - r_1)}$$

Conservation of Energy

"Energy cannot be created or destroyed... it can only be changed."

Chances are you've heard that phrase before. It's one of the most important concepts in all of physics. It doesn't mean that an object can't lose energy or gain energy. What it means is that energy can be changed into different forms, and transferred from system to system, but it never magically disappears or reappears. In the world of physics, you can never truly destroy energy. The understanding that the total amount of energy in the universe remains fixed is known as the **law of conservation of energy**.

Objects and systems can possess multiple types of energy. The energy of a system includes its kinetic energy, potential energy, and its internal energy. Mechanical energy is the sum of an object's kinetic energy as well as its gravitational potential and elastic potential energies. Non-mechanical energy forms include chemical potential, nuclear, and thermal.

Total energy is always conserved in any closed system, which is the law of conservation of energy. By confining the discussion to just the mechanical forms of energy, however, if you neglect the effects of friction you can also state that total mechanical energy is constant in any system.

Take the example of an F/A-18 Hornet jet fighter with a mass of 20,000 kilograms flying at an altitude of 10,000 meters above the surface of the earth with a velocity of 250 m/s. In this scenario, you can calculate the total mechanical energy of the jet fighter as follows by assuming ground level is gravitational potential energy level zero:

$$E_T = \Delta U_g + K = mg\Delta y + \tfrac{1}{2}mv^2$$

$$E_T = (20000\,kg)(9.8\,^m\!/_{s^2})(10000\,m) + \tfrac{1}{2}(20000\,kg)(250\,^m\!/_s)^2$$

$$E_T = 2.59 \times 10^9\,J$$

Now, assume the Hornet dives down to an altitude of 2,000 meters above the surface of the Earth. Total mechanical energy remains constant, and the gravitational potential energy of the fighter decreases, therefore the kinetic energy of the fighter must increase. The fighter's velocity goes up as a result of flying closer to the Earth! For this reason, a key concept in successful dogfighting taught to military pilots is that of energy conservation!

You can even calculate the new velocity of the fighter jet since you know its new height and its total mechanical energy must remain constant. Solving for velocity, you find that the Hornet has almost doubled its speed by "trading in" 8000 meters of altitude for velocity!

$$E_T = \Delta U_g + K = mg\Delta y + \tfrac{1}{2}mv^2$$
$$\tfrac{1}{2}mv^2 = E_T - mg\Delta y$$
$$v = \sqrt{\frac{2(E_T - mg\Delta y)}{m}}$$
$$v = \sqrt{\frac{2(2.59 \times 10^9 J - (20000 kg)(9.8 \tfrac{m}{s^2})(2000 m))}{20000 kg}}$$
$$v = 469 \tfrac{m}{s}$$

If instead you had been told that some of the mechanical energy of the jet was lost to air resistance (friction), you could also account for that by stating that the total mechanical energy of the system is equal to the gravitational potential energy, the kinetic energy, and the change in internal energy of the system (Q). This leads to the conservation of mechanical energy formula:

$$E_T = U + K + Q$$

Let's take another look at free fall, only this time, you can analyze a falling object using the law of conservation of energy and compare it to the analysis using the kinematic equations studied previously.

The problem: An object falls from a height of 10m above the ground. Neglecting air resistance, find its velocity the moment before the object strikes the ground.

Conservation of Energy Approach: The energy of the object at its highest point must equal the energy of the object at its lowest point, therefore:

$$E_{top} = E_{bottom}$$
$$U_{g_{top}} = K_{bottom}$$
$$mg\Delta y_{top} = \tfrac{1}{2}mv^2_{bottom}$$
$$v_{bottom} = \sqrt{2g\Delta y} = \sqrt{2(9.8 \tfrac{m}{s^2})(10m)} = 14 \tfrac{m}{s}$$

Kinematics Approach: For an object in free fall, its initial velocity must be zero, its displacement is 10 meters, and the acceleration due to gravity on the surface of the Earth is 9.8 m/s². Choosing down as the positive direction:

Variable	Value
v_0	0 m/s
v	FIND
Δy	10 m
a	9.8 m/s²
t	?

$$v_y^2 = v_{y0}^2 + 2a_y(y - y_0) \rightarrow v_y = \sqrt{v_{y0}^2 + 2a_y(y - y_0)} \rightarrow$$
$$v_y = \sqrt{2(9.8\,m/_{s^2})(10m)} = 14\,m/_s$$

As you can see, you reach the same conclusion regardless of approach!

5.41 Q: The diagram below shows a toy cart possessing 16 joules of kinetic energy traveling on a frictionless, horizontal surface toward a horizontal spring.

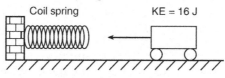

Coil spring   KE = 16 J

Frictionless, horizontal surface

If the cart comes to rest after compressing the spring a distance of 1.0 meter, what is the spring constant of the spring?

(A) 32 N/m

(B) 16 N/m

(C) 8.0 N/m

(D) 4.0 N/m

5.41 A: (A) $K = U_s = \frac{1}{2}kx^2$

$$k = \frac{2K}{x^2} = \frac{2(16J)}{(1m)^2} = 32\,N/_m$$

5.42 Q: A child does 0.20 joules of work to compress the spring in a pop-up toy. If the mass of the toy is 0.010 kilograms, what is the maximum vertical height that the toy can reach after the spring is released?

(A) 20 m

(B) 2.0 m

(C) 0.20 m

(D) 0.020 m

5.42 A: (B) The potential energy in the compressed spring must be equal to the gravitational potential energy of the toy at its maximum vertical height.

$$U_s = U_g = mg\Delta y$$

$$\Delta y = \frac{U_s}{mg} = \frac{0.2J}{(0.01kg)(9.8\,^m/_{s^2})} = 2m$$

5.43 Q: A lawyer knocks her folder of mass m off her desk of height Δy. What is the speed of the folder upon striking the floor?

(A) √(2gΔy)

(B) 2gΔy

(C) mgΔy

(D) mΔy

5.43 A: (A) The folder's initial gravitational potential energy becomes its kinetic energy right before striking the floor.

$$U_{desk} = K_{floor}$$

$$mg\Delta y = \tfrac{1}{2}mv^2$$

$$v = \sqrt{2g\Delta y}$$

5.44 Q: A 65-kilogram pole vaulter wishes to vault to a height of 5.5 meters.

(A) Calculate the minimum amount of kinetic energy the vaulter needs to reach this height if air friction is neglected and all the vaulting energy is derived from kinetic energy.

(B) Calculate the speed the vaulter must attain to have the necessary kinetic energy.

5.44 A: (A) $K = U_g = mg\Delta y$

$$K = (65kg)(9.8\,^m/_{s^2})(5.5m) = 3500J$$

(B) $K = \tfrac{1}{2}mv^2 \rightarrow v = \sqrt{\dfrac{2K}{m}} \rightarrow$

$$v = \sqrt{\frac{2(3500J)}{65kg}} = 10\,^m/_s$$

5.45 Q: The work done in accelerating an object along a frictionless horizontal surface is equal to the change in the object's

(A) momentum

(B) velocity

(C) potential energy

(D) kinetic energy

5.45 A: (D) Due to the Work-Energy Theorem.

5.46 Q: A car, initially traveling at 30 meters per second, slows uniformly as it skids to a stop after the brakes are applied. Sketch a graph showing the relationship between the kinetic energy of the car as it is being brought to a stop and the work done by friction in stopping the car.

5.46 A:

5.47 Q: A 2-kilogram block sliding down a ramp from a height of 3 meters above the ground reaches the ground with a kinetic energy of 50 joules. The total work done by friction on the block as it slides down the ramp is approximately

(A) 6 J

(B) 9 J

(C) 18 J

(D) 44 J

5.47 A: (B) The box has gravitational potential energy at the top of the ramp, which is converted to kinetic energy as it slides down the ramp. Any gravitational potential energy not converted to kinetic energy must be the work done by friction on the block, converted to internal energy (heat) of the system.

$$U_{g_{top}} = K_{bottom} + W_{friction}$$

$$W_{friction} = U_{g_{top}} - K_{bottom} = mg\Delta y - K_{bottom}$$

$$W_{friction} = (2kg)(9.8\tfrac{m}{s^2})(3m) - 50J = 9J$$

5.48 Q: Four objects travel down an inclined plane from the same height without slipping. Which will reach the bottom of the incline first?

(A) a baseball rolling down the incline

(B) an unopened soda can rolling down the incline

(C) a physics book sliding down the incline (without friction)

(D) an empty soup can rolling down the incline

5.48 A: (C) In all cases, the objects convert their gravitational potential energy into kinetic energy. In the case of the rolling objects, however, some of that kinetic energy is rotational kinetic energy. Since the physics book cannot rotate, all of its gravitational potential energy becomes translational kinetic energy; therefore, it must have the highest translational velocity.

5.49 Q: As a box is pushed 30 meters across a horizontal floor by a constant horizontal force of 25 newtons, the kinetic energy of the box increases by 300 joules. How much total internal energy is produced during this process?

(A) 150 J

(B) 250 J

(C) 450 J

(D) 750 J

5.49 A: (C). The work done on the box can be found from:

$$W = F_{||}r = (25N)(30m) = 750J$$

From the Work-Energy Theorem, you know that the total energy of the box must increase by 750 joules. If the kinetic energy of the box increases by 300 joules, where did the other 450 joules of energy go? It must have been transformed into internal energy!

5.50 Q: Mass m_1 sits on a frictionless surface and is attached by a light string across a frictionless pulley to mass m_2, as shown in the diagram below.

What happens to the gravitational potential energy and kinetic energy of m_1 and m_2 when m_2 is released from rest?

5.50 A: The gravitational potential energy of m_1 remains the same while its kinetic energy increases. The gravitational potential energy of m_2 decreases while its kinetic energy increases.

5.51 Q: A box of mass m is attached to a spring (spring constant k) and sits on a frictionless horizontal surface as shown below.

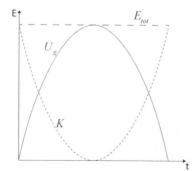

The spring is compressed a distance x from its equilibrium position and released. Determine the speed of the box when the spring returns to its equilibrium position.

5.51 A: The elastic potential energy stored in the compressed spring is converted completely to kinetic energy when the spring returns to its equilibrium position.

$$U_s = K \rightarrow \tfrac{1}{2}kx^2 = \tfrac{1}{2}mv^2 \rightarrow v = \sqrt{\frac{k}{m}}x$$

5.52 Q: A pendulum of mass M swings on a light string of length L as shown in the diagram at right. If the mass is released from rest at an angle of theta as shown, find the maximum speed of the pendulum as a function of θ, L, and any required fundamental constants.

5.52 A: First determine the gravitational potential energy at the mass's highest point as shown in problem 5.27.

$$\Delta U_g = mg\Delta y \xrightarrow{\Delta y = L(1-\cos\theta)} mgL(1 - \cos\theta)$$

Next, noting that this gravitational potential energy is converted to kinetic energy at the lowest point of the pendulum, solve for the velocity of the mass at the lowest point.

$$\Delta U_g = K \rightarrow mgL(1 - \cos\theta) = \tfrac{1}{2}mv^2 \rightarrow v = \sqrt{2gL(1 - \cos\theta)}$$

5.53 Q: A ball is thrown vertically upwards. Plot the ball's total mechanical energy, kinetic energy, and gravitational potential energy versus time on the same axes.

5.53 A:

5.54 Q: When the ball is half the distance to its peak, which of the following are true? Choose all that apply. Neglect air resistance.

(A) The ball's velocity is half its maximum velocity ($v = v_{max}/2$).

(B) The ball's velocity is its maximum velocity divided by the square root of 2 ($v = v_{max}/\sqrt{2}$).

(C) The ball's kinetic energy is equal to its gravitational potential energy.

(D) The ball's kinetic energy is equal to half its total mechanical energy.

5.54 A: (B), (C), and (D) are all true.

5.55 Q: Andy the Adventurous Adventurer, while running from evil bad guys in the Amazonian Rainforest, trips, falls, and slides down a frictionless mudslide of height 20 meters as depicted below.

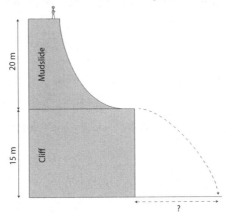

Once he reaches the bottom of the mudslide, he has the misfortune to fly horizontally off a 15-meter cliff. How far from the base of the cliff does Andy land?

5.55 A: Breaking this problem into sections, let's first analyze Andy's motion on the mudslide by utilizing conservation of energy to determine his horizontal velocity as he flies off the cliff.

$$U_g = mg\Delta y = \tfrac{1}{2}mv^2 = K \rightarrow v = \sqrt{2g\Delta y} \rightarrow$$
$$v = \sqrt{2(9.8\,\tfrac{m}{s^2})(20m)} = 19.8\,\tfrac{m}{s}$$

Once Andy flies horizontally off the cliff, this becomes a projectile problem. First find Andy's time in the air by analyzing his vertical motion, then use this time to find how far he travels horizontally before landing some distance from the base of the cliff.

$$\Delta y = v_{y0}t + \tfrac{1}{2}a_y t^2 \xrightarrow{\,v_{y0}=0\,} t = \sqrt{\frac{2\Delta y}{a_y}} = \sqrt{\frac{2(15m)}{9.8\,\tfrac{m}{s^2}}} = 1.75s$$

$$\Delta x = v_{y0}t = (19.8\,\tfrac{m}{s})(1.75s) = 34.6m$$

5.56 Q: Alicia, a 60-kg bungee jumper, steps off a 40-meter-high bridge. The bungee cord behaves like a spring with spring constant k=40 N/m. Assume there is no slack in the bungee cord.

(A) Find the speed of the jumper at a height of 15 meters above the ground.

(B) Find the speed of the jumper at a height of 30 meters above the ground.

(C) How close does the jumper get to the ground?

5.56 A: First draw a diagram of the bungee jumper, labeling the point 15 meters above the ground point A, and the point 30 meters above the ground point B.

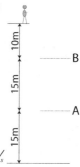

(A) Next, use conservation of energy to solve for the kinetic energy of the jumper when she is at point A.

$$U_{g_{TOP}} = U_{g_A} + U_{S_A} + K_A \rightarrow K_A = U_{g_{TOP}} - U_{g_A} - U_{S_A} \rightarrow$$

$$\tfrac{1}{2}mv_A^2 = mg\Delta y - \tfrac{1}{2}k(\Delta y)^2 \rightarrow v_A^2 = 2g\Delta y - \frac{k}{m}(\Delta y)^2 \rightarrow$$

$$v_A^2 = 2(9.8\,{}^m\!/_{s^2})(25m) - \frac{40\,{}^N\!/_m}{60kg}(25m)^2 = 73.3\,{}^{m^2}\!/_{s^2} \rightarrow v_A = 8.6\,{}^m\!/_s$$

(B) Follow the same strategy to determine the speed of the jumper at point B.

$$U_{g_{TOP}} = U_{g_B} + U_{S_B} + K_B \rightarrow K_B = U_{g_{TOP}} - U_{G_B} - U_{S_B} \rightarrow$$

$$\tfrac{1}{2}mv_B^2 = mg\Delta y - \tfrac{1}{2}k(\Delta y)^2 \rightarrow v_B^2 = 2g\Delta y - \frac{k}{m}(\Delta y)^2 \rightarrow$$

$$v_B^2 = 2(9.8\,{}^m\!/_{s^2})(10m) - \frac{40\,{}^N\!/_m}{60kg}(10m)^2 = 129\,{}^{m^2}\!/_{s^2} \rightarrow v_B = 11.4\,{}^m\!/_s$$

(C) The closest the jumper gets to the ground occurs when the jumper's speed (and therefore kinetic energy) reaches zero. Again, utilize conservation of energy.

$$U_{g_{TOP}} = U_{g_{BOTTOM}} + U_{S_{BOTTOM}} \rightarrow \Delta U_g = U_S \rightarrow mg\Delta y = \tfrac{1}{2}k(\Delta y)^2 \rightarrow$$

$$2mg = k\Delta y \rightarrow \Delta y = \frac{2mg}{k} = \frac{2(60kg)(9.8\,{}^m\!/_{s^2})}{40\,{}^N\!/_m} = 29.4m$$

If Δy=29.4m, the jumper must be 40m-29.4m from the ground at her lowest point, or 10.6m above the ground.

Chapter 5: Work, Power & Energy

Sources of Energy on Earth

So where does all this energy initially come from? Here on Earth, the energy you deal with everyday ultimately comes from the conversion of mass into energy, the source of the sun's energy. The sun's radiation provides an energy source for life on Earth, which over the millennia has become the source of fossil fuels. The sun's radiation also provides the thermal and light energy that heat the atmosphere and cause the winds to blow. The sun's energy evaporates water, which eventually recondenses as rain and snow, falling to the Earth's surface to create lakes and rivers, with gravitational potential energy, which is harnessed in hydroelectric power plants. Nuclear power also comes from the conversion of mass into energy. Just try to find an energy source on Earth that doesn't originate with the conversion of mass into energy!

Test Your Understanding

1. Design and perform an experiment which examines how a force exerted on an object does work on the object as it moves through a distance.

2. Is it possible to do work on an object without changing the object's energy? Explain why or why not with examples.

3. Humpty Dumpty falls off the wall onto the ground and breaks. If Humpty Dumpty initially has gravitational potential energy up on the wall, describe what happened to this energy, keeping in mind the law of conservation of energy.

4. Why can a single object have only kinetic energy, while potential energy requires interactions among objects or systems? Explain with real-world examples.

5. Create a conservation of energy problem that utilizes at least three different types of energy. Solve it.

6. In your own words, explain what is meant by open and closed systems. Give examples of how both energy and linear momentum are conserved in these systems.

7. An object is dropped from a given height h at the same time an identical object is launched vertically upward from ground level. Determine the initial velocity of the object launched upward such that the two objects collide at height h/2. Having already answered this problem using kinematics at the end of Chapter 3, now answer using a conservation of energy approach. Which approach is easier?

Chapter 6: Linear Momentum

"Sliding headfirst is the safest way to get to the next base, I think, and the fastest. You don't lose your momentum, and there's one more important reason I slide headfirst: it gets my picture in the paper."

— Pete Rose

Objectives

1. Define and calculate the momentum of an object.
2. Analyze situations involving average force, time, impulse, and momentum.
3. Interpret and use force vs. time graphs.
4. Apply conservation of momentum to a variety of situations.
5. Distinguish between elastic and inelastic collisions and predict the outcome of a collision based on given information.
6. Predict the velocity of the center of mass of a system when there is no interaction outside the system.

You've explored motion in some depth, specifically trying to relate what you know about motion back to kinetic energy. Recall the definition of kinetic energy as the ability or capacity of a moving object to move another object. The key characteristics of kinetic energy, mass and velocity, can be observed from the equation:

$$K = \tfrac{1}{2}mv^2$$

There's more to the story, however. Moving objects may cause other objects to move, but these interactions haven't been explored yet. To learn more about how one object causes another to move, you need to learn about collisions, and collisions are all about momentum.

Defining Momentum

Assume there's a car speeding toward you, out of control without its brakes, at a speed of 27 m/s (60 mph). Can you stop it by standing in front of it and holding out your hand? Why not?

Unless you're Superman, you probably don't want to try stopping a moving car by holding out your hand. It's too big, and it's moving way too fast. Attempting such a feat would result in a number of physics demonstrations upon your body, all of which would hurt.

You can't stop the car because it has too much momentum. **Momentum** is a vector quantity, given the symbol p, which measures how hard it is to stop a moving object. Of course, larger objects have more momentum than smaller objects, and faster objects have more momentum than slower objects. You can therefore calculate momentum using the equation:

$$\vec{p} = m\vec{v}$$

Momentum is the product of an object's mass times its velocity, and its units must be the same as the units of mass [kg] times velocity [m/s]; therefore, the units of momentum must be [kg·m/s], which can also be written as a newton-second [N·s].

6.01 Q: Two trains, Big Red and Little Blue, have the same velocity. Big Red, however, has twice the mass of Little Blue. Compare their momenta.

6.01 A: Because Big Red has twice the mass of Little Blue, Big Red must have twice the momentum of Little Blue.

6.02 Q: The magnitude of the momentum of an object is 64 kilogram-meters per second. If the velocity of the object is doubled, the magnitude of the momentum of the object will be

(A) 32 kg·m/s

(B) 64 kg·m/s

(C) 128 kg·m/s

(D) 256 kg·m/s

6.02 A: (C) if velocity is doubled, momentum is doubled.

Because momentum is a vector, the direction of the momentum vector is the same as the direction of the object's velocity.

6.03 Q: An Aichi D3A bomber, with a mass of 3600 kg, departs from its aircraft carrier with a velocity of 85 m/s due east. What is the plane's momentum?

6.03 A: $p = mv = (3600 kg)(85 \, ^m\!/_s) = 3.06 \times 10^5 \, ^{kg \bullet m}\!/_s$

Now, assume the bomber drops its payload and has burned up most of its fuel as it continues its journey east to its destination air field.

6.04 Q: If the bomber's new mass is 3,000 kg, and due to its reduced weight the pilot increases the cruising speed to 120 m/s, what is the bomber's new momentum?

6.04 A: $p = mv = (3000 kg)(120 \, ^m\!/_s) = 3.60 \times 10^5 \, ^{kg \bullet m}\!/_s$

6.05 Q: Cart A has a mass of 2 kilograms and a speed of 3 meters per second. Cart B has a mass of 3 kilograms and a speed of 2 meters per second. Compared to the inertia and magnitude of momentum of cart A, cart B has

(A) the same inertia and a smaller magnitude of momentum

(B) the same inertia and the same magnitude of momentum

(C) greater inertia and a smaller magnitude of momentum

(D) greater inertia and the same magnitude of momentum

6.05 A: (D) greater inertia and the same magnitude of momentum.

Impulse

As you can see, momentum can change, and a change in momentum is known as an **impulse**. In physics, the vector quantity impulse is represented by a capital J, and since it's a change in momentum, its units are the same as those for momentum, [kg·m/s], and can also be written as a newton-second [N·s].

$$\vec{J} = \Delta \vec{p}$$

6.06 Q: Assume the D3A bomber, which had a momentum of 3.6×10^5 kg·m/s, comes to a halt on the ground. What impulse is applied?

6.06 A: Define east as the positive direction:

$$J = \Delta p = p - p_0 = 0 - 3.6 \times 10^5 \; {}^{kg \bullet m}\!/_{s}$$

$$J = -3.6 \times 10^5 \; {}^{kg \bullet m}\!/_{s} \; \text{east} = 3.6 \times 10^5 \; {}^{kg \bullet m}\!/_{s} \; \text{west}$$

6.07 Q: Calculate the magnitude of the impulse applied to a 0.75-kilogram cart to change its velocity from 0.50 meter per second east to 2.00 meters per second east.

6.07 A: $J = \Delta p = m\Delta v = (0.75 kg)(1.5 \, {}^{m}\!/_{s}) = 1.1 N \bullet s$

6.08 Q: A 6.0-kilogram block, sliding to the east across a horizontal, frictionless surface with a momentum of 30 kilogram•meters per second, strikes an obstacle. The obstacle exerts an impulse of 10 newton•seconds to the west on the block. The speed of the block after the collision is

(A) 1.7 m/s
(B) 3.3 m/s
(C) 5.0 m/s
(D) 20 m/s

6.08 A: (B) $J = \Delta p = m\Delta v = m(v - v_0) = mv - mv_0$

$$v = \frac{J + mv_0}{m} = \frac{(-10 N \bullet s) + 30 \, {}^{kg \bullet m}\!/_{s}}{6 kg} = 3.3 \, {}^{m}\!/_{s}$$

6.09 Q: Which two quantities can be expressed using the same units?

(A) energy and force
(B) impulse and force
(C) momentum and energy
(D) impulse and momentum

6.09 A: (D) impulse and momentum both have units of kg·m/s.

6.10 Q: A 1000-kilogram car traveling due east at 15 meters per second is hit from behind and receives a forward impulse of 6000 newton-seconds. Determine the magnitude of the car's change in momentum due to this impulse.

6.10 A: Change in momentum is the definition of impulse, therefore the answer must be 6000 newton-seconds.

Impulse-Momentum Theorem

Since momentum is equal to mass times velocity, you can write that p=mv. Since you also know impulse is a change in momentum, impulse can be written as J=Δp. Combining these equations, you find:

$$J = \Delta p = \Delta mv$$

Since the mass of a single object is constant, a change in the product of mass and velocity is equivalent to the product of mass and change in velocity. Specifically:

$$J = \Delta p = m\Delta v$$

A change in velocity is called acceleration. But what causes an acceleration? A force! And does it matter if the force is applied for a very short time or a very long time? Common sense says it does and also tells us that the longer the force is applied, the longer the object will accelerate, and therefore the greater the object's change in momentum!

You can prove this using an old mathematician's trick -- if you multiply the right side of the equation by 1, you of course get the same thing. And if you multiply the right side of the equation by Δt/Δt, which is 1, you still get the same thing. Take a look:

$$J = \Delta p = \frac{m\Delta v\Delta t}{\Delta t}$$

If you look carefully at this equation, you can find a Δv/Δt, which is, by definition, acceleration. By replacing Δv/Δt with acceleration *a* in the equation, you arrive at:

$$J = \Delta p = ma\Delta t$$

One last step... perhaps you can see it already. On the right-hand side of this equation, you have ma∆t. Utilizing Newton's 2nd Law, you can replace the product of mass and acceleration with force F, giving the final form of the equation, oftentimes referred to as the **Impulse-Momentum Theorem**:

$$J = \Delta p = F\Delta t$$

This equation relates impulse to change in momentum to force applied over a time interval. For the same change in momentum, force can vary by changing the time over which it is applied. Great examples include airbags in cars, boxers rolling with punches, skydivers bending their knees upon landing, etc. To summarize, when an unbalanced force acts on an object for a period of time, a change in momentum is produced, known as an impulse.

6.11 Q: A tow-truck applies a force of 2000N on a 2000-kg car for a period of 3 seconds.

(A) What is the magnitude of the change in the car's momentum?

(B) If the car starts at rest, what will be its speed after 3 seconds?

6.11 A: (A) $\Delta p = F\Delta t = (2000\,N)(3s) = 6000\,N \bullet s$

(B) $\Delta p = p - p_0 = mv - mv_0$

$$v = \frac{\Delta p + mv_0}{m} = \frac{6000\,N \bullet s + 0}{2000kg} = 3\,m\!/\!s$$

6.12 Q: A 2-kilogram body is initially traveling at a velocity of 40 meters per second east. If a constant force of 10 newtons due east is applied to the body for 5 seconds, the final speed of the body is

(A) 15 m/s
(B) 25 m/s
(C) 65 m/s
(D) 130 m/s

6.12 A: (C) $Ft = \Delta p = m\Delta v$

$$\Delta v = v - v_0 = \frac{Ft}{m}$$

$$v = \frac{Ft}{m} + v_0$$

$$v = \frac{(10\,N)(5s)}{2kg} + 40\,m\!/\!s = 65\,m\!/\!s$$

6.13 Q: A motorcycle being driven on a dirt path hits a rock. Its 60-kilogram cyclist is projected over the handlebars at 20 meters per second into a haystack. If the cyclist is brought to rest in 0.50 seconds, the magnitude of the average force exerted on the cyclist by the haystack is

(A) 6.0×10^1 N

(B) 5.9×10^2 N

(C) 1.2×10^3 N

(D) 2.4×10^3 N

6.13 A: (D) $Ft = \Delta p = m\Delta v = m(v - v_0)$

$$F = \frac{m(v - v_0)}{t} = \frac{(60 kg)(0 - 20\,m/_s)}{0.5 s} = -2400\,N$$

6.14 Q: The instant before a batter hits a 0.14-kilogram baseball, the velocity of the ball is 45 meters per second west. The instant after the batter hits the ball, the ball's velocity is 35 meters per second east. The bat and ball are in contact for 1.0×10^{-2} second. Calculate the magnitude of the average force the bat exerts on the ball while they are in contact.

6.14 A: $Ft = \Delta p = m\Delta v$

$$F = \frac{m\Delta v}{t} = \frac{m(v - v_0)}{t}$$

$$F = \frac{(0.14 kg)(35\,m/_s - -45\,m/_s)}{1 \times 10^{-2} s} = 1120\,N$$

6.15 Q: In an automobile collision, a 44-kilogram passenger moving at 15 meters per second is brought to rest by an air bag during a 0.10-second time interval. What is the magnitude of the average force exerted on the passenger during this time?

(A) 440 N

(B) 660 N

(C) 4400 N

(D) 6600 N

6.15 A: (D) $Ft = \Delta p$

$$F = \frac{\Delta p}{t} = \frac{p - p_0}{t}$$

$$F = \frac{0 - (44 kg)(15\,m/_s)}{0.1 s} = -6600\,N$$

6.16 Q: The following carts are moving to the right across a frictionless surface with the specified initial velocity. A force is applied to each cart for a set amount of time as shown in the diagram.

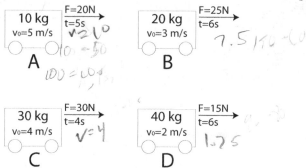

Rank the four carts from least to greatest in terms of
I) initial momentum
II) impulse applied
III) final momentum
IV) final velocity

6.16 A: I) A, B, D, C

II) D, A, C, B

III) A, D, B, C

IV) D, C, B, A

Non-Constant Forces

But not all forces are constant. What do you do if a changing force is applied for a period of time? In that case, you can make a graph of the force applied on the y-axis vs. time on the x-axis. The area under the Force-Time curve is the impulse, or change in momentum.

For the case of the sample graph at the right, you could determine the impulse applied by calculating the area of the triangle under the curve. In this case:

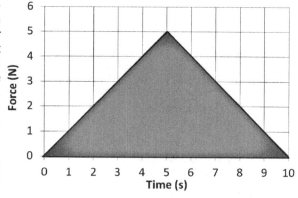

$$J = Area_{triangle} = \tfrac{1}{2}bh$$
$$J = \tfrac{1}{2}(10s)(5N) = 25N \bullet s$$

6.17 Q: A 1-kilogram box accelerates from rest in a straight line across a frictionless surface for 20 seconds as depicted in the force vs. time graph below.

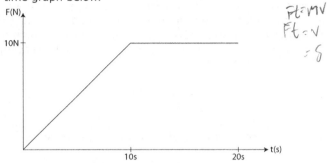

A) Find the time taken for the object to reach a speed of 5 m/s.

B) Determine the box's speed after 15 seconds.

C) What is the minimum amount of time a force of magnitude 20N could be applied to return the box to rest after its 20s acceleration?

6.17 A: A) The change in momentum, or impulse, is equal to the area under the curve. First, find the required impulse.

$$J = \Delta p = m(v - v_0) = 1kg(5\,{}^m\!/_s - 0) = 5\,{}^{kg\bullet m}\!/_s$$

The impulse is equal to the area under the graph, so find the time at which the area under the graph equals 5 N•s. Recognize that the force, in the first 10 seconds, is equal to 1 N/s × t.

$$J = Area = \tfrac{1}{2}bh \rightarrow J = \tfrac{1}{2}(t)(1t) \rightarrow J = \tfrac{1}{2}t^2 \rightarrow$$
$$t = \sqrt{2J} = \sqrt{2(5)} = \sqrt{10} = 3.16s$$

B) The change in momentum is the area under the graph from 0 to 15 seconds:

$$J = Area = m\Delta v \rightarrow Area = m(v - v_0) \xrightarrow{v_0=0} \tfrac{1}{2}bh + lw = mv \rightarrow$$
$$v = \frac{\tfrac{1}{2}bh + lw}{m} = \frac{\tfrac{1}{2}(10s)(10N) + (5s)(10N)}{1kg} = 100\,{}^m\!/_s$$

C) After 20 seconds the total momentum of the box is the area under the entire graph from 0 to 20s, or 150 N•s. The box can be stopped in the minimum amount of time by applying a 20N force in a direction opposite the box's velocity to create an impulse of -150 N•s.

$$\vec{J} = \Delta\vec{p} = \vec{F}\Delta t \rightarrow \Delta t = \frac{\vec{J}}{\vec{F}} = \frac{150N \bullet s}{20N} = 7.5s$$

Conservation of Linear Momentum

In an isolated system, where no external forces act, linear momentum is always conserved. Put more simply, in any closed system, the total momentum of the system remains constant. Therefore, an external force (an interaction with an outside object or system) is required to change the motion of an object's center of mass.

In the case of a collision or explosion (an event), if you add up the individual momentum vectors of all of the objects before the event, you'll find that they are equal to the sum of the momentum vectors of the objects after the event. Written mathematically, the law of conservation of momentum states:

$$p_{initial} = p_{final}$$

This is a direct outcome of Newton's 3rd Law.

In analyzing collisions and explosions, a momentum table can be a powerful tool for problem solving. To create a momentum table, follow these basic steps:

1. Identify all objects in the system. List them vertically down the left-hand column.
2. Determine the momenta of the objects before the event. Use variables for any unknowns.
3. Determine the momenta of the objects after the event. Use variables for any unknowns.
4. Add up all the momenta from before the event and set them equal to the momenta after the event.
5. Solve your resulting equation for any unknowns.

A **collision** is an event in which two or more objects approach and interact strongly for a brief period of time. Let's look at how the problem-solving strategy can be applied to a simple collision:

6.18 Q: A 2000-kg car traveling at 20 m/s collides with a 1000-kg car at rest at a stop sign. If the 2000-kg car has a velocity of 6.67 m/s after the collision, find the velocity of the 1000-kg car after the collision.

Chapter 6: Linear Momentum

6.18 A: Call the 2000-kg car Car A, and the 1000-kg car Car B. You can then create a momentum table as shown below:

Objects	Momentum Before (kg·m/s)	Momentum After (kg·m/s)
Car A	2000×20=40,000	2000×6.67=13,340
Car B	1000×0=0	1000×v_B=1000v_B
Total	40,000	13,340+1000v_B

Because momentum is conserved in any closed system, the total momentum before the event must be equal to the total momentum after the event.

$$40,000 = 13,340 + 1000v_B$$

$$v_B = \frac{40,000 - 13,340}{1000} = 26.7 \, ^m\!/_s$$

Not all problems are quite so simple, but problem solving steps remain consistent.

6.19 Q: On a snow-covered road, a car with a mass of 1.1×10^3 kilograms collides head-on with a van having a mass of 2.5×10^3 kilograms traveling at 8 meters per second. As a result of the collision, the vehicles lock together and immediately come to rest. Calculate the speed of the car immediately before the collision. [Neglect friction.]

6.19 A: Define the car's initial velocity as positive and the van's initial velocity as negative. After the collision, the two objects become one, therefore you can combine them in the momentum table.

Objects	Momentum Before (kg·m/s)	Momentum After (kg·m/s)
Car	1100×v_{car}=1100v_{car}	0
Van	2500×-8=-20,000	
Total	-20,000+1100v_{car}	0

$$-20000 + 1100v_{car} = 0$$

$$v_{car} = \frac{20000}{1100} = 18.2 \, ^m\!/_s$$

6.20 Q: A 70-kilogram hockey player skating east on an ice rink is hit by a 0.1-kilogram hockey puck moving toward the west. The puck exerts a 50-newton force toward the west on the player. Determine the magnitude of the force that the player exerts on the puck during this collision.

6.20 A: The player exerts a force of 50 newtons toward the east on the puck due to Newton's 3rd Law.

6.21 Q: The diagram below represents two masses before and after they collide. Before the collision, mass m_A is moving to the right with speed v, and mass m_B is at rest. Upon collision, the two masses stick together.

Before Collision   **After Collision**

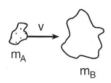

(A) $\dfrac{m_A + m_B v}{m_A}$

(C) $\dfrac{m_B v}{m_A + m_B}$

(B) $\dfrac{m_A + m_B}{m_A v}$

(D) $\dfrac{m_A v}{m_A + m_B}$

Which expression represents the speed, v', of the masses after the collision? [Assume no outside forces are acting on m_A or m_B.]

6.21 A: Use the momentum table to set up an equation utilizing conservation of momentum, then solve for the final velocity of the combined mass, labeled v'.

Objects	Momentum Before (kg·m/s)	Momentum After (kg·m/s)
Mass A	$m_A v$	$(m_A+m_B)v'$
Mass B	0	
Total	$m_A v$	$(m_A+m_B)v'$

(D) $m_A v = (m_A + m_B)v'$

$$v' = \frac{m_A v}{m_A + m_B}$$

Chapter 6: Linear Momentum

Let's take a look at another example which emphasizes the vector nature of momentum while examining an explosion. In physics terms, an **explosion** results when an object is broken up into two or more fragments.

6.22 Q: A 4-kilogram rifle fires a 20-gram bullet with a velocity of 300 m/s. Find the recoil velocity of the rifle.

6.22 A: Once again, you can use a momentum table to organize your problem-solving. To fill out the table, you must recognize that the initial momentum of the system is 0, and you can consider the rifle and bullet as a single system with a mass of 4.02 kg:

Objects	Momentum Before (kg·m/s)	Momentum After (kg·m/s)
Rifle	0	$4 \times v_{recoil}$
Bullet		$(.020)(300)=6$
Total	0	$6+4 \times v_{recoil}$

Due to conservation of momentum, you can again state that the total momentum before must equal the total momentum after, or $0=4v_{recoil}+6$. Solving for the recoil velocity of the rifle, you find:

$$0 = 4v_{recoil} + 6$$

$$v_{recoil} = \frac{-6}{4} = -1.5\,{}^{m}\!/_{s}$$

The negative recoil velocity indicates the direction of the rifle's velocity. If the bullet traveled forward at 300 m/s, the rifle must travel in the opposite direction.

6.23 Q: The diagram below shows two carts that were initially at rest on a horizontal, frictionless surface being pushed apart when a compressed spring attached to one of the carts is released. Cart A has a mass of 3.0 kilograms and cart B has a mass of 5.0 kilograms.

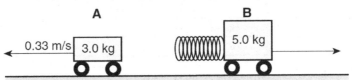

If the speed of cart A is 0.33 meter per second after the spring is released, what is the approximate speed of cart B after the spring is released?

(A) 0.12 m/s

(B) 0.20 m/s

(C) 0.33 m/s

(D) 0.55 m/s

6.23 A: Define the positive direction toward the right of the page.

Objects	Momentum Before (kg·m/s)	Momentum After (kg·m/s)
Cart A	0	3×-0.33=-1
Cart B	0	$5 \times v_B$
Total	0	$5v_B - 1$

(B) $0 = 5v_B - 1$

$$v_B = \frac{1}{5} = 0.2 \,^m\!/_s$$

6.24 Q: A woman with horizontal velocity v_1 jumps off a dock into a stationary boat. After landing in the boat, the woman and the boat move with velocity v_2. Compared to velocity v_1, velocity v_2 has

(A) the same magnitude and the same direction

(B) the same magnitude and opposite direction

(C) smaller magnitude and the same direction

(D) larger magnitude and the same direction

6.24 A: (C) due to the law of conservation of momentum.

6.25 Q: A wooden block of mass m_1 sits on a floor attached to a spring in its equilibrium position. A bullet of mass m_2 is fired with velocity v into the wooden block, where it remains. Determine the maximum displacement of the spring if the floor is frictionless and the spring has spring constant k.

6.25 A: First, find the velocity of the bullet/block system after the bullet is embedded in the block.

Objects	Momentum Before (kg·m/s)	Momentum After (kg·m/s)
Block	0	$(m_1+m_2)v'$
Bullet	m_2v	
Total	m_2v	$(m_1+m_2)v'$

$$m_2v = (m_1 + m_2)v' \rightarrow v' = \frac{m_2 v}{(m_1 + m_2)}$$

Next, recognize the kinetic energy of the bullet-block system is completely converted to elastic potential energy in the spring.

$$\tfrac{1}{2}(m_1 + m_2)v'^2 = \tfrac{1}{2}kx^2 \rightarrow x^2 = \frac{(m_1 + m_2)v'^2}{k} \xrightarrow{v' = \frac{m_2 v}{(m_1+m_2)}}$$

$$x^2 = \frac{(m_1 + m_2)}{k}\left(\frac{m_2 v}{(m_1 + m_2)}\right)^2 \rightarrow x = m_2 v\sqrt{\frac{1}{k(m_1 + m_2)}}$$

Note: It may be tempting to begin by setting the initial kinetic energy of the bullet equal to the final elastic potential energy in the spring-bullet-block system and bypass analysis of the collision. This would be incorrect, however, as the collision is inelastic. A portion of the initial kinetic energy of the bullet is transferred into internal energy of the block, which is not transformed into elastic potential energy.

6.26 Q: Two carts of differing masses are held in place by a compressed spring on a frictionless surface. When the carts are released, allowing the spring to expand, which of the following quantities will have differing magnitudes for the two cars? (Choose all that apply.)

(A) velocity

(B) acceleration

(C) force

(D) momentum

6.26 A: (A) and (B) will have differing magnitudes. Each cart will experience the same magnitude of applied force due to Newton's 3rd Law, and the magnitude of the momentum of each cart will be the same due to the law of conservation of momentum.

6.27 Q: An astronaut floating in space is motionless. The astronaut throws her wrench in one direction, propelling her in the opposite direction. Which of the following statements are true? (Choose all that apply.)

(A) The wrench will have a greater velocity than the astronaut.

(B) The astronaut will have a greater momentum than the wrench.

(C) The wrench will have greater kinetic energy than the astronaut.

(D) The astronaut will have the same kinetic energy as the wrench.

6.27 A: (A) and (C) are both true. The wrench will have greater velocity than the astronaut because the astronaut and wrench will have the same magnitude of momentum due to the law of conservation of momentum, but the wrench has a smaller mass, so must have a larger velocity. The wrench will have greater kinetic energy than the astronaut due to the kinetic energy dependence on the square of velocity.

6.28 Q: An open tub rolls across a frictionless surface. As it rolls across the surface, rain falls vertically into the tub. Which of the following statements best describe the behavior of the cart? (Choose all that apply.)

A) The tub will speed up.

B) The tub will slow down.

C) The tub's momentum will increase.

D) The tub's momentum will decrease.

6.28 A: (B) The tub will slow down. As the rain falls into the tub, the mass of the tub increases. Momentum must remain constant as no external force is applied; therefore, the velocity of the tub must decrease.

6.29 Q: A bullet of mass m_1 with velocity v_1 is fired into a block of mass m_2 attached by a string of length L in a device known as a ballistic pendulum. The ballistic pendulum records the maximum angle the string is displaced (θ). Determine the initial velocity of the bullet (v_1) in terms of m_1, m_2, L, g, and θ.

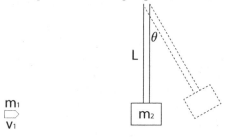

6.29 A: First determine the velocity of the bullet-block system after the collision.

Objects	Momentum Before (kg·m/s)	Momentum After (kg·m/s)
Bullet	$m_1 v$	$(m_1+m_2)v'$
Block	0	
Total	$m_1 v$	$(m_1+m_2)v'$

$$m_1 v = (m_1 + m_2)v' \rightarrow v' = \frac{m_1}{m_1 + m_2} v$$

Next, recognize that the kinetic energy of the bullet-block system immediately after the collision is converted into gravitational potential energy at the highest point of the ballistic pendulum's swing. Use this relationship to solve for the initial velocity of the bullet.

$$K_{max} = \Delta U_g \rightarrow \tfrac{1}{2}mv'^2 = mgh \rightarrow v'^2 = 2gh\xrightarrow{v' = \left(\frac{m_1}{m_1+m_2}\right)v}$$

$$\left(\frac{m_1}{m_1+m_2}\right)^2 v^2 = 2gh \rightarrow v^2 = \left(\frac{m_1+m_2}{m_1}\right)^2 2gh$$

Finally, you need to place your answer in terms of L instead of h. This can be accomplished by recognizing the geometry of the ballistic pendulum in its initial and maximum displacement positions.

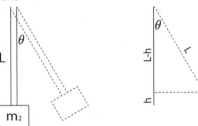

From here, an analysis of the triangle on the right allows you to solve for h in terms of L:

$$\cos\theta = \frac{adj}{hyp} \rightarrow \cos\theta = \frac{L-h}{L} \rightarrow h = L(1-\cos\theta)$$

Finally, substitute your equation for the height of the block into your equation for the velocity of the bullet to provide the initial velocity of the bullet.

$$v^2 = \left(\frac{m_1+m_2}{m_1}\right)^2 2gh\xrightarrow{h = L(1-\cos\theta)} v^2 = \left(\frac{m_1+m_2}{m_1}\right)^2 2gL(1-\cos\theta) \rightarrow$$

$$v = \left(\frac{m_1+m_2}{m_1}\right)\sqrt{2gL(1-\cos\theta)}$$

6.30 Q: A baseball and a bowling ball are rolling along a flat surface with equal momenta. How do the velocities of the balls compare?

(A) The baseball has a higher velocity than the bowling ball.

(B) The bowling ball has a higher velocity than the baseball.

(C) They are the same.

6.30 A: (A) Because they have the same momenta, and the baseball has a smaller mass, the baseball must have a higher velocity.

6.31 Q: A baseball and a bowling ball are rolling along a flat surface with equal momenta. An equal force is exerted on each to stop their motion. Which ball takes longer to come to a complete stop?

(A) The baseball takes longer to stop.

(B) The bowling ball takes longer to stop.

(C) They take the same amount of time to stop.

6.31 A: (C) They take the same amount of time to stop since an equal force is applied, and they start with the same momenta.

$$J = \Delta p = F \Delta t \rightarrow \Delta t = \frac{\Delta p}{F}$$

Since the change in momentum is the same, and the force is the same, the time the force is applied must also be the same.

6.32 Q: A 30-kg raft of dimensions 3m x 3m is motionless on a lake. A 45 kg boy crosses from one corner of the raft to the other. Neglect friction.

A) How far does the center of mass of the raft/boy system move?

B) How far does the raft move?

6.32 A: A) There is no interaction outside the system (no external force), therefore the center of mass will not move.

B) First, find the center of mass of the system in its initial state. Assume the raft's mass is centered on the raft.

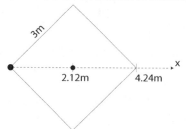

In setting up the problem, you can place the boy's path along the x-axis so that y-motion may be neglected in your analysis. Using this setup, and calling the left-most corner of the raft x=0, the boy initial starts at a position of x=0, and the raft can be modeled as a point particle of mass 30 kg at an x-position of 2.12m. The initial center of mass of the problem can then be found as:

$$x_{cm_i} = \frac{\sum m_i x_i}{\sum m_i} = \frac{(45kg)(0)+(30kg)(2.12m)}{(45kg+30kg)} = 0.848m$$

Realizing the center of mass of the system cannot move without an external force, you can then solve for the new x-position of the raft.

$$x_{cm_f} = \frac{\sum m_i x_i}{\sum m_i} = \frac{(45kg)(4.24)+(30kg)(x)}{(45kg+30kg)} = 0.848m \rightarrow x = -4.24m$$

The raft started with its center at an x-position of 2.12m, and ended with its center at an x-position of -4.24m, which means it moved 6.36m in the direction opposite the boy's displacement.

Types of Collisions

When objects collide, a number of different things can happen depending on the characteristics of the colliding objects. Of course, you know that momentum is always conserved in a closed system. Imagine, though, the differences in a collision if the two objects colliding are super-bouncy balls compared to two lumps of clay. In the first case, the balls would bounce off each other. In the second, they would stick together and become, in essence, one object. Obviously, you need more ways to characterize collisions.

Elastic collisions occur when the colliding objects bounce off of each other. This typically occurs when you have colliding objects which are very hard or bouncy. Officially, an elastic collision is one in which the sum of the kinetic energy of all the colliding objects before the event is equal to the sum of the kinetic energy of all the objects after the event. Put more simply, kinetic energy is conserved in an elastic collision.

NOTE: There is no law of conservation of kinetic energy -- IF kinetic energy is conserved in a collision, it is called an elastic collision, but there is no physical law that requires this.

Inelastic collisions occur when two objects collide and kinetic energy is not conserved. In this type of collision some of the initial kinetic energy is converted into other types of energy (heat, sound, etc.), which is why kinetic energy is NOT conserved in an inelastic collision. In a perfectly inelastic collision, the two objects colliding stick together.

In reality, most collisions fall somewhere between the extremes of a completely elastic collision and a completely inelastic collision.

6.33 Q: Two billiard balls collide. Ball 1 moves with a velocity of 4 m/s, and ball 2 is at rest. After the collision, ball 1 comes to a complete stop. What is the velocity of ball 2 after the collision? Is this collision elastic or inelastic? The mass of each ball is 0.16 kg.

6.33 A: To find the velocity of ball 2, use a momentum table.

Objects	Momentum Before (kg·m/s)	Momentum After (kg·m/s)
Ball 1	0.16×4=0.64	0
Ball 2	0	0.16×v$_2$
Total	0.64	0.16×v$_2$

$$0.64 = 0.16 \times v_2$$

$$v_2 = \frac{0.64 \frac{kg \times m}{s}}{0.16 kg} = 4 \,{}^m\!/_s$$

To determine whether this is an elastic or inelastic collision, you can calculate the total kinetic energy of the system both before and after the collision.

Objects	KE Before (J)	KE After (J)
Ball 1	0.5*0.16×4²=1.28	0
Ball 2	0	0.5*0.16×4²=1.28
Total	1.28	1.28

Since the kinetic energy before the collision is equal to the kinetic energy after the collision (kinetic energy is conserved), this is an elastic collision.

6.34 Q: Two carts of differing masses travel toward each other on a collision course as shown in the diagram below.

Cart 1   Cart 2

A) Determine the velocity of Cart 1 after the collision if Cart 2 moves to the right with a velocity of 2 m/s after the collision.

B) Is the collision elastic or inelastic?

C) If elastic, determine the kinetic energy of the system after the collision. If inelastic, identify at least one interaction that has not been considered that could account for the change in kinetic energy.

6.34 A: A) Use a momentum table to find the final velocity of Cart 1.

Objects	Momentum Before (kg·m/s)	Momentum After (kg·m/s)
Cart 1	2×2.5=5	2×v'
Cart 2	1×-1=-1	1×2
Total	4	2v'+2

Apply the law of conservation of momentum to solve for the final velocity of Cart 1.

$$4 = 2v' + 2 \rightarrow 2v' = 2 \rightarrow v' = 1\,m/s$$

B) Determine the total kinetic energy of the system before and after the collision.

Objects	K Before (J)	K After (J)
Cart 1	0.5*2×2.5²=6.25	0.5*2×1²=1
Cart 2	0.5*1×(-1)²=0.5	0.5*1×2²=2
Total	6.75	3

The kinetic energy before the collision is not equal to the kinetic energy after the collision; therefore this is an inelastic collision.

C) The change in kinetic energy could be accounted for by losses due to friction, deformation of the carts during the collision, sound, internal energy of the carts, etc.

6.35 Q: A traffic accident occurred in which a 3000-kg SUV rear-ended a 2000-kg car that was stopped at a stop sign. The diagram below depicts the skid marks of the scene after the accident.

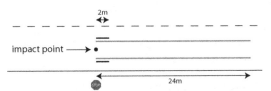

The acceleration of the car with the brakes locked is -3 m/s², and the acceleration of the SUV with the brakes locked is -2 m/s². Assuming both vehicles locked their brakes and began their skids at the instant of collision, determine the initial velocity of the truck.

6.35 A: Begin by determining the velocities of the two vehicles at the beginning of their skids (immediately after the collision) using kinematics.

Car:
$$v_x^2 = v_{x0}^2 + 2a_x(x - x_0) \rightarrow v_{x0}^2 = v_x^2 - 2a_x \Delta x \rightarrow$$
$$v_{x0}^2 = 0^2 - 2(-3)(24) \rightarrow v_{x0} = 12\,m/s$$

Truck:
$$v_{x0}^2 = 0^2 - 2(-2)(2) \rightarrow v_{x0} = 2.83\,m/s$$

Next, create a momentum table using the velocities of the two vehicles immediately after the collision.

Objects	Momentum Before (kg·m/s)	Momentum After (kg·m/s)
Car	0	2000×12=24000
Truck	3000×v=3000v	3000×(2.83)=8490
Total	3000v	32,490

Finally, apply the law of conservation of momentum to find the velocity of the truck prior to the collision.

$$3000v = 32490 \rightarrow v = 10.83 \tfrac{m}{s}$$

Collisions in 2 Dimensions

Much like the key to projectile motion, or two-dimensional kinematics problems, was breaking up vectors into their x- and y-components, the key to solving two-dimensional collision problems involves breaking up momentum vectors into x- and y- components. The law of conservation of momentum then states that momentum is independently conserved in both the x- and y- directions.

$$p_{initial_x} = p_{final_x}$$
$$p_{initial_y} = p_{final_y}$$

Therefore, you can solve two-dimensional collision problems by creating a separate momentum table for the x-component of momentum before and after the collision, and a momentum table for the y-component of momentum.

6.36 Q: Two objects of equal mass and velocities v_A and v_B collide as shown in the diagram below.

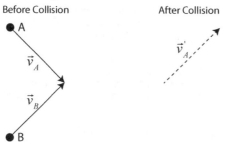

After the collision, object A travels with velocity v_A' as shown. Which vector best describes the velocity of object B after the collision?

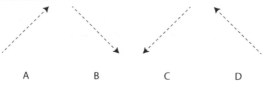

6.36 A: (B) is the only answer consistent with the law of conservation of momentum.

6.37 Q: Bert strikes a cue ball of mass 0.17 kg, giving it a velocity of 3 m/s in the x-direction. When the cue ball strikes the eight ball (mass=0.16 kg), previously at rest, the eight ball is deflected 45 degrees from the cue ball's previous path, and the cue ball is deflected 40 degrees in the opposite direction. Find the velocity of the cue ball and the eight ball after the collision.

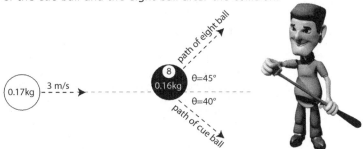

6.37 A: Start by making momentum tables for the collision, beginning with the x-direction. Since you don't know the velocity of the balls after the collision, call the velocity of the cue ball after the collision v_c, and the velocity of the eight ball after the collision v_8. Note that you must use trigonometry to determine the x-component of the momentum of each ball after the collision.

Objects	X-Momentum Before (kg·m/s)	X-Momentum After (kg·m/s)
Cue Ball	0.17×3=0.51	0.17×v_c×cos(-40°)
Eight Ball	0	0.16×v_8×cos(45°)
Total	0.51	0.17×v_c×cos(-40°)+ 0.16×v_8×cos(45°)

Since the total momentum in the x-direction before the collision must equal the total momentum in the x-direction after the collision, you can set the total before and total after columns equal:

$$0.51\,{}^{kg\bullet m}\!/_s = (0.17kg)(\cos(-40°))v_c + (0.16kg)(\cos45°)v_8$$
$$0.51\,{}^{kg\bullet m}\!/_s = (0.130kg)v_c + (0.113kg)v_8$$

Next, create a momentum table and an algebraic equation for the conservation of momentum in the y-direction.

Objects	Y-Momentum Before (kg·m/s)	Y-Momentum After (kg·m/s)
Cue Ball	0	0.17×v_c×sin(-40°)
Eight Ball	0	0.16×v_8×sin(45°)
Total	0	0.17×v_c×sin(-40°)+ 0.16×v_8×sin(45°)

$$0 = (0.17kg)(\sin-40°)v_c + (0.16kg)(\sin 45°)v_8$$
$$0 = (-0.109kg)v_c + (0.113kg)v_8$$

You now have two equations with two unknowns. To solve this system of equations, start by solving the y-momentum equation for v_c.

$$0 = (-0.109kg)v_c + (0.113kg)v_8$$
$$(0.109kg)v_c = (0.113kg)v_8$$
$$v_c = 1.04v_8$$

You can now take this equation for v_c and substitute it into the equation for conservation of momentum in the x-direction, effectively eliminating one of the unknowns, and giving a single equation with a single unknown.

$$0.51^{kg•m}\!/_s = (0.130kg)v_c + (0.113kg)v_8 \xrightarrow{v_c=1.04v_8}$$
$$0.51^{kg•m}\!/_s = (0.130kg)(1.04v_8) + (0.113kg)v_8$$
$$0.51^{kg•m}\!/_s = (0.248kg)v_8$$
$$v_8 = 2.06\,{}^{m}\!/_s$$

Finally, solve for the velocity of the cue ball after the collision by substituting the known value for v_8 into the result of the y-momentum equation.

$$v_c = 1.04v_8 \xrightarrow{v_8=2.06\,{}^{m}\!/_s}$$
$$v_c = (1.04)(2.06\,{}^{m}\!/_s) = 2.14\,{}^{m}\!/_s$$

Test Your Understanding

1. Explain Newton's 2nd Law of Motion in terms of impulse and momentum.

2. Design an experiment to test the law of conservation of momentum.

3. Explain the effect of a collision on a system's center of mass if the interacting object is part of the system. Explain the effect if the interacting object is not a part of the system.

4. Design a problem which requires application of principles of kinematics, conservation of energy, and conservation of momentum. Solve the problem you created.

5. For two objects involved in an elastic collision, prove that $v_{1i}+v_{1f}=v_{2i}+v_{2f}$.

Chapter 7: Circular Motion & Rotation

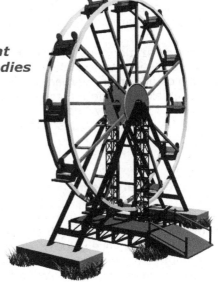

"I shall now recall to mind that the motion of the heavenly bodies is circular, since the motion appropriate to a sphere is rotation in a circle."

— Nicolaus Copernicus

Objectives

1. Explain the acceleration of an object moving in a circle at constant speed.
2. Define centripetal force and recognize that it is not a special kind of force, but that it is provided by forces such as tension, gravity, and friction.
3. Solve problems involving calculations of centripetal force.
4. Calculate the period, frequency, speed and distance traveled for objects moving in circles at constant speed.
5. Differentiate between translational and rotational motion of an object.
6. Describe the rotational motion of an object in terms of rotational position, velocity, and acceleration.
7. Use rotational kinematic equations to solve problems for objects rotating at constant acceleration.
8. Utilize the definitions of torque and Newton's 2nd Law for Rotational Motion to solve static equilibrium problems.
9. Apply principles of angular momentum, torque, and rotational dynamics to analyze a variety of situations.
10. Explain what is meant by conservation of angular momentum.
11. Calculate the rotational kinetic energy and total kinetic energy of a rotating object moving through space.

Now that you've talked about linear and projectile kinematics, as well as fundamentals of dynamics and Newton's Laws, you have the skills and background to analyze circular motion. Of course, this has obvious applications such as cars moving around a circular track, roller coasters moving in loops, and toy trains chugging around a circular track under the Christmas tree. Less obvious, however, is the application to turning objects. Any object that is turning can be thought of as moving through a portion of a circular path, even if it's just a small portion of that path.

With this realization, analysis of circular motion will allow you to explore a car speeding around a corner on an icy road, a running back cutting upfield to avoid a blitzing linebacker, and the orbits of planetary bodies. The key to understanding all of these phenomena starts with the study of uniform circular motion.

Uniform Circular Motion

The motion of an object in a circular path at constant speed is known as **uniform circular motion (UCM)**. An object in UCM is constantly changing direction, and since velocity is a vector and has direction, you could say that an object undergoing UCM has a constantly changing velocity, even if its speed remains constant. And if the velocity of an object is changing, it must be accelerating. Therefore, an object undergoing UCM is constantly accelerating. This type of acceleration is known as **centripetal acceleration**.

7.01 Q: If a car is accelerating, is its speed increasing?

7.01 A: It depends. Its speed could be increasing, or it could be accelerating in a direction opposite its velocity (slowing down). Or, its speed could remain constant yet still be accelerating if it is traveling in uniform circular motion.

Just as importantly, you'll need to figure out the direction of the object's acceleration, since acceleration is a vector. To do this, draw an object moving counter-clockwise in a circular path, and show its velocity vector at two different points in time. Since acceleration is the rate of change of an object's velocity with respect to time, you can determine the direction of the object's acceleration by finding the direction of its change in velocity, Δv.

To find its change in velocity, Δv, recall that $\Delta v = v - v_0$.

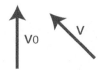

Therefore, you can find the difference of the vectors v and v_0 graphically, which can be re-written as $\Delta v = v + (-v_0)$.

Recall that to add vectors graphically, you line them up tip-to-tail, then draw the resultant vector from the starting point (tail) of the first vector to the ending point (tip) of the last vector.

So, the acceleration vector must point in the direction shown above. If this vector is shown back on the original circle, lined up directly between the initial and final velocity vector, it's easy to see that the acceleration vector points toward the center of the circle.

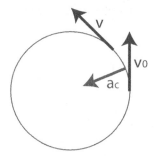

You can repeat this procedure from any point on the circle... no matter where you go, the acceleration vector always points toward the center of the circle. In fact, the word centripetal in centripetal acceleration means "center-seeking!"

So now that you know the direction of an object's acceleration (toward the center of the circle), what about its magnitude? The formula for the magnitude of an object's centripetal acceleration is given by:

$$a_c = \frac{v^2}{r}$$

7.02 Q: In the diagram below, a cart travels clockwise at constant speed in a horizontal circle.

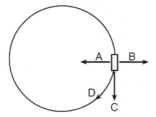

At the position shown in the diagram, which arrow indicates the direction of the centripetal acceleration of the cart?

(A) A

(B) B

(C) C

(D) D

7.02 A: (A) The acceleration of any object moving in a circular path is toward the center of the circle.

7.03 Q: The diagram shows the top view of a 65-kilogram student at point A on an amusement park ride. The ride spins the student in a horizontal circle of radius 2.5 meters, at a constant speed of 8.6 meters per second. The floor is lowered and the student remains against the wall without falling to the floor.

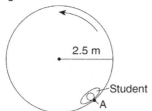

Which vector best represents the direction of the centripetal acceleration of the student at point A?

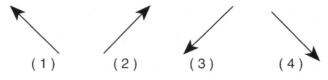

7.03 A: (1) Centripetal acceleration points toward the center of the circle.

7.04 Q: Which graph best represents the relationship between the magnitude of the centripetal acceleration and the speed of an object moving in a circle of constant radius?

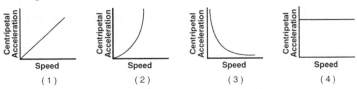

7.04 A: (2) Centripetal acceleration is proportional to v²/r.

7.05 Q: A car rounds a horizontal curve of constant radius at a constant speed. Which diagram best represents the directions of both the car's velocity, v, and acceleration, a?

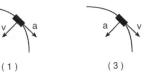

7.05 A: (3) Velocity is tangent to the circular path, and acceleration is toward the center of the circular path.

7.06 Q: A 0.50-kilogram object moves in a horizontal circular path with a radius of 0.25 meter at a constant speed of 4.0 meters per second. What is the magnitude of the object's acceleration?

(A) 8 m/s²

(B) 16 m/s²

(C) 32 m/s²

(D) 64 m/s²

7.06 A: (D) 64 m/s².

Circular Speed

So how do you find the speed of an object as it travels in a circular path? The formula for speed that you learned in kinematics still applies.

$$\bar{v} = \frac{d}{t}$$

You have to be careful in using this equation, however, to understand that an object traveling in a circular path is traveling along the circumference of a circle. Therefore, if an object were to make one complete revolution around the circle, the distance it travels is equal to the circle's circumference.

$$C = 2\pi r$$

7.07 Q: Miranda drives her car clockwise around a circular track of radius 30m. She completes 10 laps around the track in 2 minutes. Find Miranda's total distance traveled, average speed, and centripetal acceleration.

7.07 A: $d = 10 \times 2\pi r = 10 \times 2\pi(30m) = 1885m$

$$\overline{v} = \frac{d}{t} = \frac{1885m}{120s} = 15.7 \, ^m\!/_s$$

$$a_c = \frac{v^2}{r} = \frac{(15.7\,^m\!/_s)^2}{30m} = 8.2 \, ^m\!/_{s^2}$$

Note that her displacement and average velocity are zero.

7.08 Q: The combined mass of a race car and its driver is 600 kilograms. Traveling at constant speed, the car completes one lap around a circular track of radius 160 meters in 36 seconds. Calculate the speed of the car.

7.08 A: $\overline{v} = \dfrac{d}{t} = \dfrac{2\pi r}{t} = \dfrac{2\pi(160m)}{36s} = 27.9 \, ^m\!/_s$

Centripetal Force

If an object traveling in a circular path has an inward acceleration, Newton's 2nd Law states there must be a net force directed toward the center of the circle as well. This type of force, known as a **centripetal force**, can be a gravitational force, a tension, an applied force, or even a frictional force.

NOTE: When dealing with circular motion problems, it is important to realize that a centripetal force isn't really a new force, a centripetal force is just a label or grouping you apply to a force to indicate its direction is toward the center of a circle. This means that you **never** want to label a force on a free body diagram as a centripetal force, F_c. Instead, label the center-directed force as specifically as you can. If a tension is causing the force, label the force F_T. If a frictional force is causing the center-directed force, label it F_f, and so forth. Because a centripetal force is always perpendicular to the object's motion, a centripetal force can do no work on an object.

You can combine the equation for centripetal acceleration with Newton's 2nd Law to obtain Newton's 2nd Law for Circular Motion. Recall that Newton's 2nd Law states:

$$F_{net} = ma$$

For an object traveling in a circular path, there must be a net (centripetal) force directed toward the center of the circular path to cause a (centripetal) acceleration directed toward the center of the circular path. You can revise Newton's 2nd Law for this particular case as follows:

$$F_C = ma_C$$

Then, recalling the formula for centripetal acceleration as:

$$a_c = \frac{v^2}{r}$$

You can put these together, replacing a_c in the equation to get a combined form of Newton's 2nd Law for Uniform Circular Motion:

$$F_C = \frac{mv^2}{r}$$

Of course, if an object is traveling in a circular path and the centripetal force is removed, the object will continue traveling in a straight line in whatever direction it was moving at the instant the force was removed.

7.09 Q: An 800N running back turns a corner in a circular path of radius 1 meter at a velocity of 8 m/s. Find the running back's mass, centripetal acceleration, and centripetal force.

7.09 A: $mg = 800N \qquad m = \dfrac{800N}{9.8 \, ^m\!/_{s^2}} = 81.5kg$

$a_c = \dfrac{v^2}{r} = \dfrac{(8 \, ^m\!/_s)^2}{1m} = 64 \, ^m\!/_{s^2}$

$F_c = ma_c = (81.5kg)(64 \, ^m\!/_{s^2}) = 5220N$

7.10 Q: The diagram at right shows a 5.0-kilogram bucket of water being swung in a horizontal circle of 0.70-meter radius at a constant speed of 2.0 meters per second. The magnitude of the centripetal force on the bucket of water is approximately

(A) 5.7 N

(B) 14 N

(C) 29 N

(D) 200 N

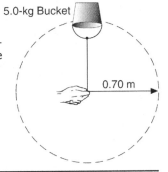

5.0-kg Bucket

0.70 m

7.10 A: (C) $F_c = ma_c = m\dfrac{v^2}{r} = (5kg)\dfrac{(2\,{}^m\!/_s)^2}{0.7m} = 29\,N$

7.11 Q: A 1.0×10^3-kilogram car travels at a constant speed of 20 meters per second around a horizontal circular track. Which diagram correctly represents the direction of the car's velocity (v) and the direction of the centripetal force (F_c) acting on the car at one particular moment?

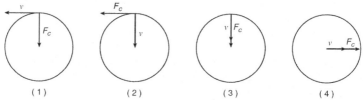

(1)   (2)   (3)   (4)

7.11 A: (1) Velocity is tangent to the circle, and the centripetal force points toward the center of the circle. Note that these are NOT FBDs, because they show more than forces acting on a single object.

7.12 Q: A 1750-kilogram car travels at a constant speed of 15 meters per second around a horizontal, circular track with a radius of 45 meters. The magnitude of the centripetal force acting on the car is

(A) 5.00 N

(B) 583 N

(C) 8750 N

(D) 3.94 10^5 N

7.12 A: (C) $F_c = ma_c = m\dfrac{v^2}{r} = (1750kg)\dfrac{(15\,{}^m\!/_s)^2}{45m} = 8750\,N$

7.13 Q: A ball attached to a string is moved at constant speed in a horizontal circular path. A target is located near the path of the ball as shown in the diagram. At which point along the ball's path should the string be released, if the ball is to hit the target?

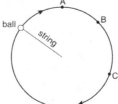

(A) A

(B) B

(C) C

(D) D

7.13 A: (B) If released at point B, the ball will continue in a straight line to the target.

7.14 Q: A 1200-kilogram car traveling at a constant speed of 9 meters per second turns at an intersection. The car follows a horizontal circular path with a radius of 25 meters to point P. At point P, the car hits an area of ice and loses all frictional force on its tires. Which path does the car follow on the ice? (Choose A, B, C, or D)

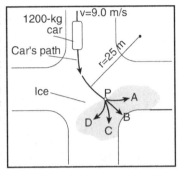

7.14 A: (B) Once the car loses all frictional force, there is no longer a force toward the center of the circular path, therefore the car will travel in a straight line toward B.

Frequency and Period

For objects moving in circular paths, you can characterize their motion around the circle using the terms frequency (f) and period (T). The **frequency** of an object is the number of revolutions the object makes in a complete second. It is measured in units of [1/s], or Hertz (Hz). In similar fashion, the **period** of an object is the time it takes to make one complete revolution. Since the period is a time interval, it is measured in units of seconds. You can relate period and frequency using the equations:

$$f = \frac{1}{T} \qquad T = \frac{1}{f}$$

7.15 Q: A 500g toy train completes 10 laps of its circular track in 1 minute and 40 seconds. If the diameter of the track is 1 meter, find the train's centripetal acceleration (a_c), centripetal force (F_c), period (T), and frequency (f).

7.15 A: $\bar{v} = \dfrac{d}{t} = \dfrac{2\pi r \times 10}{t} = \dfrac{2\pi(0.5m) \times 10}{100s} = 0.314\,{}^{m}\!/_{s}$

$a_c = \dfrac{v^2}{r} = \dfrac{(0.314\,{}^{m}\!/_{s})^2}{0.5m} = 0.197\,{}^{m}\!/_{s^2}$

$F_C = ma_c = (0.5kg)(0.197\,{}^{m}\!/_{s^2}) = 0.099\,N$

$T = \dfrac{100s}{10revs} = 10s$

$f = \dfrac{1}{T} = \dfrac{1}{10s} = 0.1Hz$

7.16 Q: Alan makes 38 complete revolutions on the playground Round-A-Bout in 30 seconds. If the radius of the Round-A-Bout is 1 meter, determine

(A) Period of the motion

(B) Frequency of the motion

(C) Speed at which Alan revolves

(D) Centripetal force on 40-kg Alan

7.16 A: (A) $T = \dfrac{30s}{38revs} = 0.789s$

(B) $f = \dfrac{1}{T} = \dfrac{1}{0.789s} = 1.27\,Hz$

(C) $\bar{v} = \dfrac{d}{t} = \dfrac{38 \times 2\pi r}{t} = \dfrac{38 \times 2\pi(1m)}{30s} = 7.96\,{}^{m}\!/_{s}$

(D) $F_C = ma_c = m\dfrac{v^2}{r} = (40kg)\dfrac{(7.96\,{}^{m}\!/_{s})^2}{1m} = 2530\,N$

Vertical Circular Motion

Objects travel in circles vertically as well as horizontally. Because the speed of these objects isn't typically constant, technically this isn't uniform circular motion, but your UCM analysis skills still prove applicable.

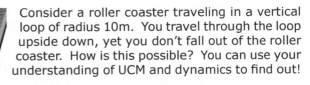

Consider a roller coaster traveling in a vertical loop of radius 10m. You travel through the loop upside down, yet you don't fall out of the roller coaster. How is this possible? You can use your understanding of UCM and dynamics to find out!

To begin with, first take a look at the coaster when the car is at the bottom of the loop. Drawing a free body diagram, the force of gravity on the coaster, also known as its weight, pulls it down; so draw a vector pointing down labeled "mg." Opposing that force is the normal force of the rails of the coaster pushing up, which is labeled F_N.

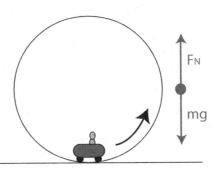

Because the coaster is moving in a circular path, you can analyze it using the tools developed for uniform circular motion. Newton's 2nd Law still applies, so you can write:

$$F_{NET_C} = F_N - mg$$

Notice that because you're talking about circular motion, you can adopt the convention that forces pointing toward the center of the circle are positive, and forces pointing away from the center of the circle are negative. At this point, recall that the force you "feel" when you're in motion is actually the normal force. So, solving for the normal force as you begin to move in a circle, you find:

$$F_N = F_{NET_C} + mg$$

Since you know that the net force is always equal to mass times acceleration, the net centripetal force is equal to mass times the centripetal acceleration. You can therefore replace F_{NETc} as follows:

$$F_N = F_{NET_C} + mg = \frac{mv^2}{r} + mg$$

You can see from the resulting equation that the normal force is now equal to the weight plus an additional term from the centripetal force of the circular motion. As you travel in a circular path near the bottom of the loop, you feel heavier than your weight. In common terms, you feel additional "g-forces." How many g's you feel can be obtained with a little bit more manipulation. If you re-write your equation for the normal force, pulling out the mass by applying the distributive property of multiplication, you obtain:

$$F_N = m\left(\frac{v^2}{r} + g\right)$$

Notice that inside the parenthesis you have the standard acceleration due to gravity, g, plus a term from the centripetal acceleration. This additional term is the additional g-force felt by a person. For example, if a_c was equal to g (9.8 m/s²), you could say the person in the cart was experiencing two g's (1g from the centripetal acceleration, and 1g from the Earth's gravitational field). If a_c were equal to 3×g (29.4 m/s²), the person would be experiencing a total of four g's.

Expanding this analysis to a similar situation in a different context, try to imagine instead of a roller coaster, a mass whirling in a vertical circle by a string. You could replace the normal force by the tension in the string in the analysis. Because the force is larger at the bottom of the circle, the likelihood of the string breaking is highest when the mass is at the bottom of the circle!

At the top of the loop, you have a considerably different picture. Now, the normal force from the coaster rails must be pushing down against the cart, though still in the positive direction since down is now toward the center of the circular path. In this case, however, the weight of the object also points toward the center of the circle, since the Earth's gravitational field always pulls toward the center of the Earth. The Free Body Diagram looks considerably different, and therefore the application of Newton's 2nd Law for Circular Motion is considerably different as well:

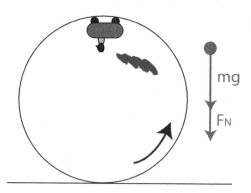

$$F_{NET_C} = F_N + mg$$

Since the force you feel is actually the normal force, you can solve for the normal force and expand the net centripetal force as shown:

$$F_N = F_{NET_C} - mg = \frac{mv^2}{r} - mg$$

You can see from the equation that the normal force is now the centripetal force minus your weight. If the centripetal force were equal to your weight, you would feel as though you were weightless. Note that this is also the point where the normal force is exactly equal to 0. This means the rails of the track are no longer pushing on the roller coaster cart. If the centripetal force was slightly smaller, and the cart's speed was slightly smaller, the normal force F_N would be less than 0. Since the rails can't physically pull the cart in the negative direction (away from the center of the circle), this means the car is falling off the rail and the cart's occupant is about to have a very, very bad day. Only by maintaining a high speed can the cart successfully negotiate the loop. Go too slow and the cart falls.

In order to remain safe, real roller coasters actually have wheels on both sides of the rails to prevent the cart from falling if it ever did slow down at the top of a loop, although coasters are designed so that this situation never actually occurs.

7.17 Q: In an experiment, a rubber stopper is attached to one end of a string that is passed through a plastic tube before weights are attached to the other end. The stopper is whirled in a horizontal circular path at constant speed.

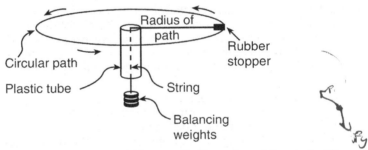

(A) Describe what would happen to the radius of the circle if the student whirls the stopper at a greater speed without changing the balancing weights.

(B) The rubber stopper is now whirled in a vertical circle at the same speed. On the vertical diagram, draw and label vectors to indicate the direction of the weight (F_g) and the direction of the centripetal force (F_c) at the position shown.

7.17 A: (A) As the speed of the stopper is increased, the radius of the orbit will increase.

(B)

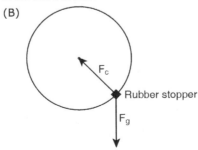

Rotational Kinematics

The motion of objects cannot always be described completely using the laws of physics that you've looked at so far. Besides motion, which changes an object's overall position (translational motion, or translational displacement), many objects also rotate around an axis, known as rotational, or angular, motion. The motion of some objects involves both translational and rotational motion.

An arrow speeding to its target, a hovercraft maneuvering through a swamp, and a hot air balloon floating through the atmosphere all experience only translational motion. A Ferris Wheel at an amusement park, a top spinning on a table, and a carousel at the beach experience only rotational motion. The Earth rotating around its axis (rotational motion) and moving through space (translational motion), and a frisbee spinning around its center while also flying through the air, both demonstrate simultaneous translational and rotational motion.

Typically people discuss rotational motion in terms of degrees, where one entire rotation around a circle is equal to 360°. When dealing with rotational motion from a physics perspective, measuring rotational motion in units known as radians (rads) is much more efficient. A **radian** measures a distance around an arc equal to the length of the arc's radius.

Up to this point, you've described distances and displacements in terms of Δx and Δy. In discussing **angular displacements**, you must transition to describing the translational displacement around an arc in terms of the variable s, while continuing to use the symbol θ (theta) to represent angles and angular displacement.

The distance completely around a circular path (360°), known as the **circumference**, C, can be found using $\Delta s = C = 2\pi r = 2\pi$ radians. Therefore, you can use this as a conversion factor to move back and forth between degrees and radians.

7.18 Q: Convert 90° to radians.

7.18 A: $90° \times \dfrac{2\pi \text{ rads}}{360°} = \dfrac{\pi}{2} \text{ rads} = 1.57 \text{ rads}$

7.19 Q: Convert 6 radians to degrees.

7.19 A: $6 \text{ rads} \times \dfrac{360°}{2\pi \text{ rads}} = 344°$

Angles are also measured in terms of revolutions (complete trips around a circle). A complete single rotation is equal to 360°; therefore you can write the conversion factors for rotational distances and displacements as $360° = 2\pi$ radians $= 1$ revolution.

7.20 Q: Convert 1.5 revolutions to both radians and degrees.

7.20 A: $1.5 \text{ revs} \times \dfrac{2\pi \text{ rad}}{1 \text{ rev}} = 3\pi \text{ rads}$

$1.5 \text{ revs} \times \dfrac{360°}{1 \text{ rev}} = 540°$

Rotational kinematics is extremely similar to translational kinematics: all you have to do is learn the rotational versions of the kinematic variables and equations. When you learned translation kinematics, displacement was discussed in terms of Δx. With rotational kinematics, you'll use the angular coordinate θ instead. When average velocity was introduced in the translational world, you used the formula:

$$\bar{v} = \frac{x - x_0}{t} = \frac{\Delta x}{t}$$

When exploring rotational motion, you'll talk about the **angular velocity** ω (omega), given in units of radians per second (rad/s). Because angular (rotational) velocity is a vector, define the positive direction of rotation as counter-clockwise around the circular path, and the negative direction as clockwise around the path.

$$\bar{\omega} = \frac{\theta - \theta_0}{t} = \frac{\Delta \theta}{t}$$

NOTE: Formally, the direction of angular vectors is determined by the right-hand rule. Wrap the fingers of your right hand in the direction of the rotational displacement, velocity, or acceleration, and your thumb will point in the vector's direction.

7.21 Q: A record spins on a phonograph at 33 rpms (revolutions per minute) clockwise. Find the angular velocity of the record.

7.21 A:

$$\bar{\omega} = \frac{\Delta\theta}{t} = \frac{-33 \text{ revs}}{1\min}$$

$$\frac{-33 \text{ revs}}{1\min} \times \frac{1\min}{60s} \times \frac{2\pi \text{ rad}}{1 \text{ rev}} = -3.46 \, ^{rad}\!/_{s}$$

Note that the angular velocity vector is negative because the record is rotating in a clockwise direction.

7.22 Q: Find the magnitude of Earth's angular velocity in radians per second.

7.22 A: Realizing that the Earth makes one complete revolution every 24 hours, we can estimate the magnitude of the Earth's angular velocity as:

$$\bar{\omega} = \frac{\Delta\theta}{t} = \frac{2\pi \text{ rads}}{24 \text{ hr}} \times \frac{1 \text{ hr}}{3600s} = 7.27 \times 10^{-5} \, ^{rads}\!/_{s}$$

In similar fashion, when you learned about translational acceleration, you found acceleration as the rate of change of an object's translational velocity:

$$a = \frac{\Delta v}{t} = \frac{v - v_0}{t}$$

Angular acceleration α (alpha), given in units of radians per second^2, is the rate of change of an object's angular velocity. Since angular acceleration is a vector as well, you can define its direction as positive for increasing angular velocities in the counter-clockwise direction, and negative for increasing angular velocities in the clockwise direction.

$$\alpha = \frac{\Delta\omega}{t}$$

7.23 Q: A frog rides a unicycle. If the unicycle wheel begins at rest, and accelerates uniformly in a counter-clockwise direction to an angular velocity of 15 rpms in a time of 6 seconds, find the angular acceleration of the unicycle wheel.

7.23 A: First, convert 15 rpms to rads/s.

$$\frac{15 \text{ revs}}{\text{min}} \times \frac{1 \text{ min}}{60s} \times \frac{2\pi \text{ rad}}{1 \text{ rev}} = 1.57 \, ^{rad}\!/\!_s$$

Next, use the definition of angular acceleration.

$$\alpha = \frac{\Delta\omega}{t} = \frac{\omega - \omega_0}{t} = \frac{(1.57 \, ^{rad}\!/\!_s)\text{-}0}{6s} = 0.26 \, ^{rad}\!/\!_{s^2}$$

Again, note the positive angular acceleration, as the bicycle wheel is accelerating in the counter-clockwise direction.

Putting these definitions together, you observe a very strong parallel between translational kinematic quantities and rotational kinematic quantities.

Variable	Translational	Angular
Displacement	Δs	Δθ
Velocity	v	ω
Acceleration	a	α
Time	t	t

It's quite straightforward to translate between translational and angular variables as well when you know the radius (*r*) of the point of interest on a rotating object and assume the object is not "slipping" as it rotates.

Variable	Translational	Angular
Displacement	$s = r\theta$	$\theta = \dfrac{s}{r}$
Velocity	$v = r\omega$	$\omega = \dfrac{v}{r}$
Acceleration	$a = r\alpha$	$\alpha = \dfrac{a}{r}$
Time	t	t

7.24 Q: A knight swings a mace of radius 1m in two complete revolutions. What is the distance traveled by the mace?

(A) 3.1 m

(B) 6.3 m

(C) 12.6 m

(D) 720 m

7.24 A: (C) $s = r\theta = (1m)(4\pi \text{ rads}) = 12.6m$

7.25 Q: A compact disc player is designed to vary the disc's rotational velocity so that the point being read by the laser moves at a linear velocity of 1.25 m/s. What is the CD's rotational velocity in revs/s when the laser is reading information on an inner portion of the disc at a radius of 0.03m?

7.25 A: $\omega = \dfrac{v}{r} = \dfrac{1.25\,^m/_s}{0.03m} = 41.7\,^{rad}/_s$

$$\dfrac{41.7 \text{ rad}}{s} \times \dfrac{1 \text{ rev}}{2\pi \text{ rad}} = 6.63\,^{rev}/_s$$

7.26 Q: What is the rotational velocity of the compact disc in the previous problem when the laser is reading the outermost portion of the disc (radius=0.06m)?

7.26 A: $\omega = \dfrac{v}{r} = \dfrac{1.25\,^m/_s}{0.06m} = 20.8\,^{rad}/_s$

$$\dfrac{20.8 \text{ rad}}{s} \times \dfrac{1 \text{ rev}}{2\pi \text{ rad}} = 3.32\,^{rev}/_s$$

7.27 Q: A carousel accelerates from rest to an angular velocity of 0.3 rad/s in a time of 10 seconds. What is its angular acceleration? What is the linear acceleration for a point at the outer edge of the carousel, at a radius of 2.5 meters from the axis of rotation?

7.27 A: $\alpha = \dfrac{\omega - \omega_0}{t} = \dfrac{0.3\,^{rad}/_s}{10s} = 0.03\,^{rad}/_{s^2}$

$a = r\alpha = (2.5m)(0.03\,^{rad}/_{s^2}) = 0.075\,^{m}/_{s^2}$

The parallels between translational and rotational motion go even further. You developed a set of kinematic equations for translational motion that allowed you to explore the relationship between displacement, velocity, and acceleration. You can develop a corresponding set of relationships for angular displacement, angular velocity, and angular acceleration. The equations follow the same form as the translational equations; all you have to do is replace the translational variables with rotational variables, as shown in the following table.

Translational	Rotational
$v_x = v_{x0} + a_x t$	$\omega = \omega_0 + \alpha t$
$x = x_0 + v_{x0}t + \frac{1}{2}a_x t^2$	$\theta = \theta_0 + \omega_0 t + \frac{1}{2}\alpha t^2$
$v_x^2 = v_{x0}^2 + 2a_x(x - x_0)$	$\omega^2 = \omega_0^2 + 2\alpha(\theta - \theta_0)$

The rotational kinematic equations can be used the same way you used the translational kinematic equations to solve problems. Once you know three of the kinematic variables, you can always use the equations to solve for the other two.

7.28 Q: A carpenter cuts a piece of wood with a high powered circular saw. The saw blade accelerates from rest with an angular acceleration of 14 rad/s² to a maximum speed of 15,000 rpms. What is the maximum speed of the saw in radians per second?

7.28 A: $$\frac{15{,}000\ \text{revs}}{\text{min}} \times \frac{1\ \text{min}}{60s} \times \frac{2\pi\ \text{rad}}{1\ \text{rev}} = 1570\ ^{rad}\!/_s$$

7.29 Q: How long does it take the saw to reach its maximum speed?

7.29 A: You can use the rotational kinematic equations to solve this problem:

Variable	Value
ω_0	0 rad/s
ω	1570 rad/s
$\Delta\theta$	?
α	14 rad/s²
t	FIND

$$\omega = \omega_0 + \alpha t$$

$$t = \frac{\omega - \omega_0}{\alpha} = \frac{1570\ ^{rad}\!/_s - 0}{14\ ^{rad}\!/_{s^2}} = 112s$$

7.30 Q: How many complete rotations does the saw make while accelerating to its maximum speed?

7.30 A: $$\theta = \theta_0 + \omega_0 t + \tfrac{1}{2}\alpha t^2$$

$$\theta = \tfrac{1}{2}(14\ ^{rad}\!/_{s^2})(112s)^2 = 87{,}800\ \text{rads}$$

$$87{,}800\ \text{rad} \times \frac{1\ \text{rev}}{2\pi\ \text{rads}} = 14{,}000\ \text{revolutions}$$

Chapter 7: Circular Motion & Rotation

7.31 Q: A safety mechanism will bring the saw blade to rest in 0.3 seconds should the carpenter's hand come off the saw controls. What angular acceleration does this require? How many complete revolutions will the saw blade make in this time?

7.31 A: Begin by re-creating the rotational kinematics table.

Variable	Value
ω_0	1570 rad/s
ω	0 rad/s
$\Delta\theta$	FIND
α	FIND
t	0.3s

First, find the angular acceleration.

$$\omega = \omega_0 + \alpha t$$

$$\alpha = \frac{\omega - \omega_0}{t} = \frac{0 - 1570\,^{rad}/_s}{0.3s} = -5230\,^{rad}/_{s^2}$$

Next, find the angular displacement.

$$\Delta\theta = \omega_0 t + \tfrac{1}{2}\alpha t^2$$

$$\Delta\theta = (1570\,^{rad}/_s)(0.3s) + \tfrac{1}{2}(-5230\,^{rad}/_{s^2})(0.3s)^2$$

$$\Delta\theta = 236 \text{ rads}$$

Finally, convert the angular displacement into revolutions

$$\Delta\theta = 236 \text{ rads} \times \frac{1 \text{ rev}}{2\pi \text{ rads}} = 37.5 \text{ revolutions}$$

7.32 Q: An amusement park ride of radius x allows children to sit in a spinning swing held by a cable of length L.

Photo Courtesy of Wolfgang Sauber

At maximum angular speed, the cable makes an angle of θ with the vertical as shown in the diagram below. Determine the maximum angular speed of the rider in terms of g, θ, x and L.

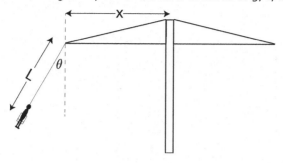

7.32 A: First write a Newton's 2nd Law equation for the x-direction, calling the tension in the cable T. Note that since the rider is moving in a circular path, the net force in the x-direction must also be the rider's centripetal force.

$$F_{net_x} = T \sin \theta = ma = \frac{mv^2}{r}$$

Then, write a Newton's 2nd Law equation for the y-direction.

$$F_{net_y} = T \cos \theta - mg = 0 \rightarrow T \cos \theta = mg$$

Now notice that you can divide the first equation by the second equation to eliminate T.

$$\frac{T \sin \theta}{T \cos \theta} = \frac{\frac{mv^2}{r}}{mg} = \frac{mv^2}{mgr} \rightarrow \tan \theta = \frac{v^2}{gr}$$

Next, convert translational speed to angular speed and solve for angular speed ω.

$$\tan \theta = \frac{v^2}{gr} \xrightarrow{v=\omega r} \tan \theta = \frac{\omega^2 r}{g} \rightarrow \omega = \sqrt{\frac{g \tan \theta}{r}}$$

Finally, format your answer in terms of g, θ, x and L by replacing r with x+Lsinθ.

$$\omega = \sqrt{\frac{g \tan \theta}{r}} \xrightarrow{r=x+L\sin\theta} \omega = \sqrt{\frac{g \tan \theta}{x + L \sin \theta}}$$

Torque

Torque (τ) is a force that causes an object to turn. If you think about using a wrench to tighten a bolt, the closer to the bolt you apply the force, the harder it is to turn the wrench, while the farther from the bolt you apply the force, the easier it is to turn the wrench. This is because you generate a larger torque when you apply a force at a greater distance from the axis of rotation.

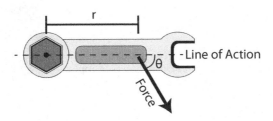

Let's take a look at the example of a wrench turning a bolt. A force is applied at a distance from the axis of rotation. Call this distance r. When you apply forces at 90 degrees to the imaginary line leading from the axis of rotation to the point where the force is applied (known as the **line of action**), you obtain maximum torque. As the angle at which the force applied decreases (θ), so does the torque causing the bolt to turn. Therefore, you can calculate the torque applied as:

$$\tau = Fr\sin\theta = r_{\perp}F = rF\sin\theta$$

In some cases, physicists will refer to $r\sin\theta$ (or $r_{\perp}$) as the **lever arm**, or **moment arm**, of the system. The lever arm is the perpendicular distance from the axis of rotation to the point where the force is applied. Alternately, you could think of torque as the component of the force perpendicular to the lever multiplied by the distance r. Units of torque are the units of force × distance, or newton-meters (N·m).

Objects which have no rotational acceleration, or a net torque of zero, are said be in **rotational equilibrium**. This implies that any net positive (counter-clockwise) torque is balanced by an equal net negative (clockwise) torque, and the object will remain in its current state of rotation.

7.33 Q: A pirate captain takes the helm and turns the wheel of his ship by applying a force of 20 Newtons to a wheel spoke. If he applies the force at a radius of 0.2 meters from the axis of rotation, at an angle of 80° to the line of action, what torque does he apply to the wheel?

7.33 A: $\tau = rF\sin\theta$

$\tau = (0.2m)(20N)\sin(80°) = 3.94N \bullet m$

7.34 Q: A mechanic tightens the lugs on a tire by applying a torque of 110 N·m at an angle of 90° to the line of action. What force is applied if the wrench is 0.4 meters long?

7.34 A: $\tau = rF \sin \theta$

$$F = \frac{\tau}{r \sin \theta} = \frac{110N \bullet m}{(0.4m)\sin 90°} = 275N$$

7.35 Q: What is the minimum length of the wrench if the mechanic is only capable of applying a force of 200N?

7.35 A: $\tau = rF \sin \theta$

$$r = \frac{\tau}{F \sin \theta} = \frac{110N \bullet m}{(200N)\sin 90°} = .55m$$

7.36 Q: A constant force F is applied for five seconds at various points of the object below, as shown in the diagram. Rank the magnitude of the torque exerted by the force on the object about an axle located at the center of mass from smallest to largest.

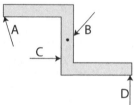

7.36 A: B, C, A, D

7.37 Q: A variety of masses are attached at different points to a uniform beam attached to a pivot. Rank the angular acceleration of the beam from largest to smallest.

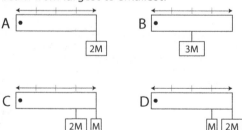

7.37 A: D, C, A, B

7.38 Q: A 3-kilogram café sign is hung from a 1-kilogram horizontal pole as shown in the diagram. A wire is attached to prevent the sign from rotating. Find the tension in the wire.

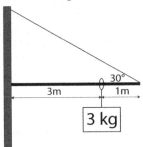

7.38 A: Start by drawing a diagram of the horizontal pole, showing all the forces on the pole as a means to illustrate the various torques. Assume the pivot is the attachment point on the left hand side of the pole.

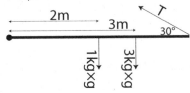

Since the pole is in equilibrium, the net torque must be zero.

$$\tau_{net} = T\sin 30°(4m) - (3kg)(g)(3m) - (1kg)(g)(2m) = 0 \rightarrow$$

$$T = \frac{11\ kg \bullet m \times g}{4m\sin 30°} = 54N$$

7.39 Q: A 10-kg tortoise sits on a see-saw 1 meter from the fulcrum. Where must a 2-kg hare sit in order to maintain rotational equilibrium? Assume the see-saw is massless.

7.39 A: First draw a diagram of the situation, showing the forces on the see-saw.

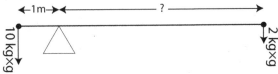

Since the see-saw is in rotational equilibrium, the net torque must be zero.

$$\tau_{net} = 0 \rightarrow (10kg)(g)(1m) - (2kg)(g)(x) = 0 \rightarrow x = 5m$$

Moment of Inertia

Previously, the inertial mass of an object (its translational inertia) was defined as that object's ability to resist a linear acceleration. Similarly, an object's rotational inertia, or **moment of inertia**, describes an object's resistance to a rotational acceleration. Objects that have most of their mass near their axis of rotation have a small rotational inertia, while objects that have more mass farther from the axis of rotation have larger rotational inertias. The symbol for an object's moment of inertia is I.

For common objects, you can look up the formula for their moment of inertia. For more complex objects, the moment of inertia can be calculated by taking the sum of all the individual particles of mass making up the object multiplied by the square of their radius from the axis of rotation.

$$I = \sum mr^2$$

This can be quite cumbersome using algebra for continuous objects, and is therefore typically left to calculus-based courses or numerical approximation using computing systems.

Commonly Used Moments of Inertia

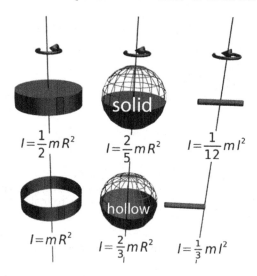

$$I = \frac{1}{2}mR^2 \qquad I = \frac{2}{5}mR^2 \qquad I = \frac{1}{12}ml^2$$

$$I = mR^2 \qquad I = \frac{2}{3}mR^2 \qquad I = \frac{1}{3}ml^2$$

7.40 Q: Calculate the moment of inertia for a solid sphere with a mass of 10 kg and a radius of 0.2m.

7.40 A: $I = \frac{2}{5}mR^2$

$I = \frac{2}{5}(10kg)(0.2m)^2$

$I = 0.16kg \bullet m^2$

7.41 Q: Find the moment of inertia (I) of two 5-kg bowling balls joined by a meter-long rod of negligible mass when rotated about the center of the rod. Compare this to the moment of inertia of the object when rotated about one of the masses.

7.41 A: $I = \sum mr^2 = (5kg)(0.5m)^2 + (5kg)(0.5m)^2 = 2.5kg \bullet m^2$

$I = \sum mr^2 = (5kg)(1m)^2 + (5kg)(0)^2 = 5kg \bullet m^2$

7.42 Q: Calculate the moment of inertia for a hollow sphere with a mass of 10 kg and a radius of 0.2 m.

7.42 A: $I = \frac{2}{3}mR^2$

$I = \frac{2}{3}(10kg)(0.2m)^2$

$I = 0.27kg \bullet m^2$

7.43 Q: Calculate the moment of inertia for a long thin rod with a mass of 2 kg and a length of 1m rotating around the center of its length.

7.43 A: $I = \frac{1}{12}ml^2$

$I = \frac{1}{12}(2kg)(1m)^2$

$I = 0.17kg \bullet m^2$

7.44 Q: Calculate the moment of inertia for a long thin road with a mass of 2kg and a length of 1m rotating about its end.

7.44 A: $I = \frac{1}{3}ml^2$

$I = \frac{1}{3}(2kg)(1m)^2$

$I = 0.67kg \bullet m^2$

7.45 Q: An object with uniform mass density is rotated about an axle, which may be in position A, B, C, or D. Rank the object's moment of inertia from smallest to largest based on axle position.

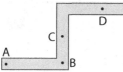

7.45 A: C, B, D, A

Newton's 2nd Law for Rotation

In the chapter on dynamics, you learned about forces causing objects to accelerate. The larger the net force, the greater the linear (or translational) acceleration, and the larger the mass of the object, the smaller the translational acceleration.

$$F_{net} = ma$$

The rotational equivalent of this law, Newton's 2nd Law for Rotation, relates the torque on an object to its resulting angular acceleration. The larger the net torque, the greater the rotational acceleration, and the larger the rotational inertia, the smaller the rotational acceleration:

$$\tau_{net} = I\alpha$$

7.46 Q: What is the angular acceleration experienced by a uniform solid disc of mass 2 kg and radius 0.1 m when a net torque of 10 N·m is applied? Assume the disc spins about its center.

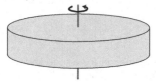

7.46 A: $\tau_{net} = I\alpha = \frac{1}{2}mR^2\alpha$

$$\alpha = \frac{2\tau_{net}}{mR^2} = \frac{2 \times 10N \bullet m}{(2kg)(0.1m)^2} = 1000 \, {}^{rad}\!/\!{}_{s^2}$$

7.47 Q: A Round-A-Bout on a playground with a moment of inertia of 100 kg·m² starts at rest and is accelerated by a force of 150N at a radius of 1m from its center. If this force is applied at an angle of 90° from the line of action for a time of 0.5 seconds, what is the final rotational velocity of the Round-A-Bout?

7.47 A: Start by making our rotational kinematics table:

Variable	Value
ω_0	0 rad/s
ω	FIND
$\Delta\theta$	?
α	?
t	0.5s

Since you only know two items on the table, you must find a third before you solve this with the rotational kinematic equations. Since you are given the moment of inertia of the Round-A-Bout as well as the applied force, you can solve for the angular acceleration using Newton's 2nd Law for Rotational Motion.

$$\tau_{net} = I\alpha$$

$$\alpha = \frac{\tau_{net}}{I} = \frac{Fr\sin\theta}{I} = \frac{(150N)(1m)\sin 90°}{100kg \bullet m^2} = 1.5\,{}^{rad}\!/\!{}_{s^2}$$

Now, use your rotational kinematics to solve for the final angular velocity of the Round-A-Bout.

$$\omega = \omega_0 + \alpha t$$

$$\omega = 0 + (1.5\,{}^{rad}\!/\!{}_{s^2})(0.5s) = 0.75\,{}^{rad}\!/\!{}_{s}$$

7.48 Q: A top with moment of inertia 0.001 kg×m² is spun on a table by applying a torque of 0.01 N×m for two seconds. If the top starts from rest, find the final angular velocity of the top.

7.48 A: First find the angular acceleration using Newton's 2nd Law for Rotation.

$$\tau_{net} = I\alpha \rightarrow \alpha = \frac{\tau_{net}}{I} = \frac{0.01N \bullet m}{0.001kg \bullet m^2} = 10\,{}^{rad}\!/\!{}_{s^2}$$

Next, use rotational kinematics to find the top's final angular velocity.

$$\omega = \omega_0 + \alpha t = 0 + (10\,{}^{rad}\!/\!{}_{s^2})(2s) = 20\,{}^{rad}\!/\!{}_{s}$$

7.49 Q: A light string attached to a mass m is wrapped around a pulley of mass m_p and radius R. If the moment of the inertia of the pulley is $\frac{1}{2}m_pR^2$, find the acceleration of the mass.

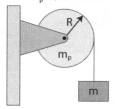

7.49 A: First draw Free Body Diagrams for both the pulley as well as the hanging mass.

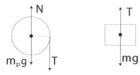

Write a Newton's 2nd Law for Rotation equation for the pulley and solve for the tension in the string.

$$\tau_{net} = I\alpha \xrightarrow{\tau=rF\sin\theta}_{I=\frac{1}{2}m_pR^2} RT = \frac{1}{2}m_pR^2\alpha \longrightarrow$$

$$T = \frac{1}{2}m_pR\alpha \xrightarrow{a=R\alpha} T = \frac{1}{2}m_pa$$

Next, write a Newton's 2nd Law Equation in the y-direction for the hanging mass. Note that you can substitute in your value for T derived from the pulley FBD to solve for the acceleration.

$$mg - T = ma \xrightarrow{T=\frac{1}{2}m_pa} mg - \frac{1}{2}m_pa = ma \longrightarrow$$

$$mg = a\left(m + \frac{m_p}{2}\right) \longrightarrow a = \frac{mg}{m + \frac{m_p}{2}}$$

7.50 Q: Two blocks are connected by a light string over a pulley of mass m_p and radius R as shown in the diagram below.

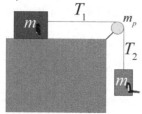

Find the acceleration of mass m_2 if m_1 sits on a frictionless surface.

7.50 A: First draw Free Body Diagrams for the masses and the pulley.

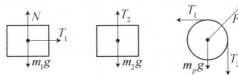

Start by writing Newton's 2nd Law equations for the x-direction for m_1, the y-direction for m_2, and rotation for the pulley. Note that T is not constant due to the pulley having inertia.

$$F_{net_x} = T_1 = m_1 a$$

$$F_{net_y} = m_2 g - T_2 = m_2 a \rightarrow T_2 = m_2 g - m_2 a$$

$$\tau_{net} = T_2 R - T_1 R = I\alpha$$

Next combine the equations to eliminate the tensions, then solve for the acceleration using the relationship between angular acceleration and translational acceleration. Assume the pulley is a uniform disc.

$$\tau_{net} = T_2 R - T_1 R = I\alpha \xrightarrow[T_2 = m_2 g - m_2 a]{T_1 = m_1 a} R(m_2 g - m_2 a - m_1 a) = I\alpha \xrightarrow[a = \alpha R]{I_{disc} = \frac{1}{2} m_p R^2}$$

$$R(m_2 g - m_2 a - m_1 a) = \tfrac{1}{2} m_p R^2 \left(\frac{a}{R}\right) = \tfrac{1}{2} m_p a R \rightarrow m_2 g - m_2 a - m_1 a = \tfrac{1}{2} m_p a \rightarrow$$

$$m_2 g = m_1 a + m_2 a + \tfrac{1}{2} m_p a \rightarrow m_2 g = a(m_1 + m_2 + \tfrac{1}{2} m_p) \rightarrow a = \frac{m_2 g}{m_1 + m_2 + \tfrac{1}{2} m_p}$$

Angular Momentum

Much like linear momentum describes how difficult it is to stop an object moving linearly, **angular momentum** (L) is a vector describing how difficult it is to stop a rotating object. The total angular momentum of a system is the sum of the individual angular momenta of the objects comprising that system.

A mass with velocity **v** moving at some position **r** about point Q has angular momentum **L_Q**, as shown in the diagram at right. Note that angular momentum depends on the point of reference.

$$\left|\vec{L}_Q\right| = mvr\sin\theta \xrightarrow{v = \omega r} \left|\vec{L}_Q\right| = mr^2\omega$$

For an object rotating about its center of mass, the angular momentum of the object is equal to the product of the object's moment of inertia and its angular velocity.

$$L = I\omega$$

This is an intrinsic property of an object rotating about its center of mass, and is known as the object's spin angular momentum. It is constant even if you calculate it relative to any point in space.

Previously, you learned that linear momentum, the product of an object's mass and its velocity, is conserved in a closed system. In similar fashion, **spin angular momentum** L, the product of an object's moment of inertia and its angular velocity about the center of mass, is also conserved in a closed system with no external net torques applied. This is the law of **conservation of angular momentum**.

This can be observed by watching a spinning ice skater. As an ice skater launches into a spin, she generates rotational velocity by applying a torque to her body. The skater now has an angular momentum as she spins around an axis which is equal to the product of her moment of inertia (rotational inertia) and her rotational velocity.

To increase the rotational velocity of her spin, she pulls her arms in close to her body, reducing her moment of inertia. Angular momentum is conserved, therefore rotational velocity must increase. Then, before coming out of the spin, the skater reduces her rotational velocity by move her arms away from her body, increasing her moment of inertia.

When a net force is applied to an object for a specific amount of time, the object receives an impulse, which causes a change in linear momentum. Similarly, when a net torque is applied to an object for a specific amount of time, the object receives an angular impulse, which causes a change in angular momentum.

$$\Delta \vec{p} = \vec{F}\Delta t$$
$$\Delta \vec{L} = \vec{\tau}\Delta t$$

7.51 Q: Find the angular momentum for a 5 kg point particle located at (2,2) with a velocity of 2 m/s east

A) about the origin (point O)

B) about point P at (2,0)

C) about point Q at (0,2)

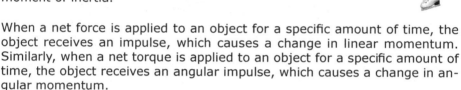

7.51 A: A) $\left|\vec{L}_O\right| = mvr\sin\theta = (5)(2)(2\sqrt{2})\sin 45° = 20\,^{kg \bullet m^2}\!/_s$

B) $\left|\vec{L}_P\right| = mvr\sin\theta = (5)(2)(2)\sin 90° = 20\,^{kg \bullet m^2}\!/_s$

C) $\left|\vec{L}_Q\right| = mvr\sin\theta = (5)(2)(2)\sin 0° = 0$

7.52 Q: Angelina spins on a rotating pedestal with an angular velocity of 8 radians per second. Bob throws her an exercise ball, which increases her moment of inertia from 2 kg·m² to 2.5 kg·m². What is Angelina's angular velocity after catching the exercise ball? (Neglect any external torque from the ball.)

7.52 A: Since there are no external torques, you know that the initial spin angular momentum must equal the final spin angular momentum, and can therefore solve for Angelina's final angular velocity:

$$L_0 = L \rightarrow I_0 \omega_0 = I\omega$$

$$\omega = \frac{I_0 \omega_0}{I} = \frac{(2kg \bullet m^2)(8\,{}^{rad}\!/_s)}{2.5kg \bullet m^2} = 6.4\,{}^{rad}\!/_s$$

7.53 Q: A disc with moment of inertia 1 kg·m² spins about an axle through its center of mass with angular velocity 10 rad/s. An identical disc which is not rotating is slid along the axle until it makes contact with the first disc. If the two discs stick together, what is their combined angular velocity?

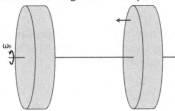

7.53 A: Once again, since there are no external torques, you know that spin angular momentum must be conserved. When the two discs stick together, their new combined moment of inertia must be the sum of their individual moments of inertia, for a total moment of inertia of 2 kg·m².

$$L_0 = L \rightarrow I_0 \omega_0 = I\omega$$

$$\omega = \frac{I_0 \omega_0}{I} = \frac{(1kg \bullet m^2)(10\,{}^{rad}\!/_s)}{2kg \bullet m^2} = 5\,{}^{rad}\!/_s$$

7.54 Q: A constant force F is applied for a constant time at various points of the object below, as shown in the diagram. Rank the magnitude of the change in the object's angular momentum due to the force from smallest to largest.

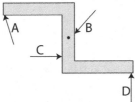

7.54 A: B, C, A, D

Rotational Kinetic Energy

When kinematics was first introduced, kinetic energy was defined as the ability of a moving object to move another object. Then, translational kinetic energy for a moving object was calculated using the formula:

$$K = \tfrac{1}{2}mv^2$$

Since an object which is rotating also has the ability to move another object, it, too, must have kinetic energy. Rotational kinetic energy can be calculated using the analog to the translational kinetic energy formula -- all you have to do is replace inertial mass with moment of inertia, and translational velocity with angular velocity!

$$K_{rot} = \tfrac{1}{2}I\omega^2$$

If an object exhibits both translational motion and rotational motion, the total kinetic energy of the object can be found by adding the translational kinetic energy and the rotational kinetic energy:

$$K = \tfrac{1}{2}mv^2 + \tfrac{1}{2}I\omega^2$$

Because you're solving for energy, of course the answers will have units of Joules.

7.55 Q: Gina rolls a bowling ball of mass 7 kg and radius 10.9 cm down a lane with a velocity of 6 m/s. Find the rotational kinetic energy of the bowling ball, assuming it does not slip.

7.55 A: To find the rotational kinetic energy of the bowling ball, you need to know its moment of inertia and its angular velocity. Assume the bowling ball is a solid sphere to find its moment of inertia.

$$I = \tfrac{2}{5}mR^2$$
$$I = \tfrac{2}{5}(7kg)(.109m)^2 = 0.0333kg \bullet m^2$$

Next, find the ball's angular velocity.

$$\omega = \frac{v}{r} = \frac{6\,{}^m\!/_s}{.109m} = 55\,{}^{rad}\!/_s$$

Finally, solve for the rotational kinetic energy of the bowling ball.

$$K_{rot} = \tfrac{1}{2}I\omega^2 = \tfrac{1}{2}(0.0333kg \bullet m^2)(55\,{}^{rad}\!/_s)^2 = 50.4J$$

7.56 Q: Find the total kinetic energy of the bowling ball from the previous problem.

7.56 A: The total kinetic energy is the sum of the translational kinetic energy and the rotational kinetic energy of the bowling ball.

$$K = \tfrac{1}{2}mv^2 + \tfrac{1}{2}I\omega^2$$
$$K = \tfrac{1}{2}(7kg)(6\,{}^m\!/_s)^2 + 50.4J$$
$$K = 176J$$

7.57 Q: Harrison kicks a soccer ball which rolls across a field with a velocity of 5 m/s. What is the ball's total kinetic energy? You may assume the ball doesn't slip, and treat it as a hollow sphere of mass 0.43 kg and radius 0.11 meter.

7.57 A: Immediately note that the ball will have both translational and rotational kinetic energy. Therefore, you'll need to know the ball's mass (given), translational velocity (given), moment of inertia (unknown), and rotational velocity (unknown).

Start by finding the moment of inertia of the ball, modeled as a hollow sphere.

$$I = \tfrac{2}{3}mR^2$$
$$I = \tfrac{2}{3}(0.43kg)(0.11m)^2 = 0.00347\,kg \bullet m^2$$

Next, find the rotational velocity of the soccer ball.

$$\omega = \frac{v}{r} = \frac{5\,{}^m\!/_s}{0.11m} = 45.5\,{}^{rad}\!/_s$$

Now you have enough information to calculate the total kinetic energy of the soccer ball.

$$K = \tfrac{1}{2}mv^2 + \tfrac{1}{2}I\omega^2$$
$$K = \tfrac{1}{2}(0.43kg)(5\,{}^m\!/_s)^2 + \tfrac{1}{2}(0.00347\,kg \bullet m^2)(45.5\,{}^{rad}\!/_s)^2$$
$$K = 8.96J$$

7.58 Q: An ice skater spins with a specific angular velocity. She brings her arms and legs closer to her body, reducing her moment of inertia to half its original value. What happens to her angular velocity? What happens to her rotational kinetic energy?

7.58 A: As the skater pulls her arms and legs in, she reduces her moment of inertia. Since there is no external net torque, her spin angular momentum remains constant; therefore her angular velocity must double. Rotational kinetic energy, on the other hand, is governed by $K=\frac{1}{2}I\omega^2$. Moment of inertia is cut in half, but angular velocity is doubled; therefore rotational kinetic energy is doubled. The skater does work in pulling her arms and legs in while spinning.

7.59 Q: Find the speed of a disc of radius R which starts at rest and rolls without slipping down an incline of height H.

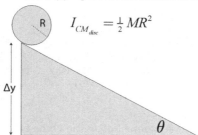

$$I_{CM_{disc}} = \tfrac{1}{2}MR^2$$

7.59 A: Use a conservation of energy approach, recognizing that the gravitational potential energy of the disc at the top of incline must equal the kinetic energy of the disc at the bottom of the incline.

$$K_i + U_i = K_f + U_f \xrightarrow[\substack{U_f=0}]{K_i=0} U_i = K_f$$

Next, recognize that the kinetic energy has translational and rotational components, then solve for the final velocity of the center of mass, utilizing the relationship between angular velocity and translational velocity.

$$Mg\Delta y = \tfrac{1}{2}Mv_{CM}^2 + \tfrac{1}{2}I\omega^2 = \tfrac{1}{2}Mv_{CM}^2 + \tfrac{1}{2}(\tfrac{1}{2}MR^2)\omega^2 \rightarrow$$

$$g\Delta y = \tfrac{1}{2}v_{CM}^2 + \tfrac{1}{4}R^2\omega^2 \xrightarrow{v=\omega R} g\Delta y = \tfrac{1}{2}v_{CM}^2 + \tfrac{1}{4}v_{CM}^2 \rightarrow$$

$$g\Delta y = \tfrac{3}{4}v_{CM}^2 \rightarrow v_{CM} = \sqrt{\tfrac{4}{3}g\Delta y}$$

Putting all this information together, rotational physics mirrors translational physics in terms of both variables and formulas. These equivalencies and relationships are summarized below.

Variable	Translational	Angular
Displacement	Δs	$\Delta \theta$
Velocity	v	ω
Acceleration	a	α
Time	t	t
Force/Torque	F	τ
Mass/Moment of Inertia	m	I

Variable	Translational	Angular
Displacement	$s = r\theta$	$\theta = \dfrac{s}{r}$
Velocity	$v = r\omega$	$\omega = \dfrac{v}{r}$
Acceleration	$a = r\alpha$	$\alpha = \dfrac{a}{r}$
Time	t	t
Force/Torque	$F_{net} = ma$	$\tau_{net} = I\alpha$
Momentum	$p = mv$	$L = I\omega$
Impulse	$\Delta p = F\Delta t$	$\Delta L = \tau \Delta t$
Kinetic Energy	$K = \frac{1}{2}mv^2$	$K = \frac{1}{2}I\omega^2$

Test Your Understanding

1. Explain Newton's 2nd Law of Motion in terms of impulse and momentum. Compare your understanding to Newton's 2nd Law of Motion for Rotation in terms of angular impulse and angular momentum.

2. A fly hangs on to the edge of a Frisbee as it flies through the air. Describe the forces acting on the fly. Draw a vector diagram showing the fly's angular velocity, translational velocity, and total velocity.

3. Consider the right-hand-rule and how it can be used to determine the direction of the angular velocity, angular acceleration, and angular momentum vectors. Does this make sense? Why or why not?

4. Using conservation of angular momentum, explain why it is harder to remain upright on a bicycle at slower speeds, and easier to remain upright when the bicycle is moving at faster speeds.

5. A marble is rolled separately down two different inclines of the same height as shown below. Compare the speed of the marble at the bottom of incline A to the speed of the marble at the bottom of incline B. Compare the time it takes the marble to reach the bottom of incline A to the time it takes to reach the bottom of incline B. Justify your answers.

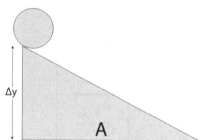 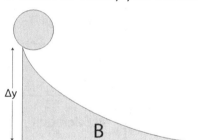

Chapter 8: Gravity

"I can calculate the motion of heavenly bodies, but not the madness of people."

— Sir Isaac Newton

Objectives

1. Calculate the gravitational force on an object in a gravitational field.
2. Utilize Newton's Law of Universal Gravitation to determine the gravitational force of attraction between two objects.
3. Determine the gravitational field due to an object.
4. Differentiate between inertial mass and gravitational mass.
5. Calculate the gravitational potential energy of a system.
6. Explain weightlessness for objects in orbit.
7. Describe key characteristics of geosynchronous and geostationary orbits.
8. Explain how Kepler's Laws describe the orbits of planetary objects around the sun.

Universal Gravitation

All objects that have mass attract each other with a gravitational force. The magnitude of that force, F_g, can be calculated using Newton's Law of Universal Gravitation:

$$\left|F_g\right| = G\frac{m_1 m_2}{r^2}$$

This law says that the force of gravity between two objects is proportional to each of the masses(m_1 and m_2) and inversely proportional to the square of the distance between them (r). The **universal gravitational constant**, G, is a "fudge factor," so to speak, included in the equation so that your answers come out in S.I. units. G is given as 6.67×10^{-11} N·m²/kg².

Let's look at this relationship in a bit more detail. Force is directly proportional to the masses of the two objects. Therefore, if either of the masses were doubled, the gravitational force would also double. In similar fashion, if the distance between the two objects, r, was doubled, the force of gravity would be quartered since the distance is squared in the denominator. This type of relationship is called an inverse square law, which describes many phenomena in the natural world.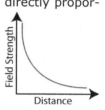

NOTE: The distance between the masses, r, is actually the distance between the center of masses of the objects. For large objects, such as the gravitational attraction between the Earth and the moon, you must determine the distance from the center of the Earth to the center of the moon, not their surfaces.

Back in chapter 1, you learned that inertial mass is an object's resistance to being accelerated by a force, and more massive objects accelerate less than smaller objects given an identical force. You also learned that gravitational mass relates to the amount of gravitational force experienced by an object: objects with larger gravitational mass experience a larger gravitational force. Thankfully, inertial mass and gravitational mass have been tested over and over again and are always exactly the same.

Some hints for problem solving when dealing with Universal Gravitation:

1. Substitute values in for variables at the very end of the problem only. The longer you can keep the formula in terms of variables, the fewer opportunities for mistakes.
2. Before using your calculator to find an answer, try to estimate the order of magnitude of the answer. Use this to check your final answer.
3. Once your calculations are complete, make sure your answer makes sense by comparing your answer to a known or similar quantity. If your answer doesn't make sense, check your work and verify your calculations.

8.1 Q: What is the magnitude of the gravitational force of attraction between two asteroids in space, each with a mass of 50,000 kg, separated by a distance of 3800 m?

8.1 A:
$$\left|F_g\right| = G\frac{m_1 m_2}{r^2}$$

$$\left|F_g\right| = (6.67\times10^{-11}\ ^{N\bullet m^2}/_{kg^2})\frac{(50000kg)(50000kg)}{(3800m)^2} = 1.15\times10^{-8}\ N$$

As you can see, the force of gravity is a relatively weak force, and you would expect a relatively weak force between relatively small objects. It takes tremendous masses and relatively small distances in order to develop significant gravitational forces. Let's take a look at another problem to explore the relationship between gravitational force, mass, and distance.

8.2 Q: As a meteor moves from a distance of 16 Earth radii to a distance of 2 Earth radii from the center of Earth, the magnitude of the gravitational force between the meteor and Earth becomes

(A) one-eighth as great

(B) 8 times as great

(C) 64 times as great

(D) 4 times as great

8.2 A: (C) 64 times as great. The gravitational force is given by Newton's Law of Universal Gravitation. If the radius is one-eighth its initial value, and radius is squared in the denominator, the radius squared becomes one-sixty fourth its initial value. Because radius squared is in the denominator, the gravitational force must increase by 64X.

8.3 Q: Which diagram best represents the gravitational forces, F_g, between a satellite, S, and Earth?

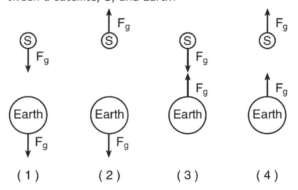

(1)　　　(2)　　　(3)　　　(4)

8.3 A: (3) Newton's 3rd Law says that the force of gravity on the satellite due to Earth will be equal in magnitude and opposite in the direction the force of gravity on the Earth due to the satellite.

8.4 Q: Io (pronounced "EYE oh") is one of Jupiter's moons discovered by Galileo. Io is slightly larger than Earth's Moon. The mass of Io is 8.93×10^{22} kilograms and the mass of Jupiter is 1.90×10^{27} kilograms. The distance between the centers of Io and Jupiter is 4.22×10^8 meters.

(A) Calculate the magnitude of the gravitational force of attraction that Jupiter exerts on Io.

(B) Calculate the magnitude of the acceleration of Io due to the gravitational force exerted by Jupiter.

8.4 A: (A) 6.35×10^{22} N

$$\left| F_g \right| = G \frac{m_1 m_2}{r^2} = (6.67 \times 10^{-11}\ {}^{N \bullet m^2}\!/_{kg^2}) \frac{(8.93 \times 10^{22}\ kg)(1.9 \times 10^{27}\ kg)}{(4.22 \times 10^8\ m)^2}$$

$$\left| F_g \right| = 6.35 \times 10^{22}\ N$$

(B) $a = \dfrac{F_{net}}{m} = \dfrac{6.35 \times 10^{22}\ N}{8.93 \times 10^{22}\ kg} = 0.71\ {}^{m}\!/_{s^2}$

8.5 Q: The diagram shows two bowling balls, A and B, each having a mass of 7 kilograms, placed 2 meters apart.

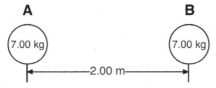

What is the magnitude of the gravitational force exerted by ball A on ball B?

(A) 8.17×10^{-9} N

(B) 1.63×10^{-9} N

(C) 8.17×10^{-10} N

(D) 1.17×10^{-10} N

8.5 A: (C) $\left| F_g \right| = G \dfrac{m_1 m_2}{r^2} = (6.67 \times 10^{-11}\ {}^{N \bullet m^2}\!/_{kg^2}) \dfrac{(7\,kg)(7\,kg)}{(2m)^2} = 8.17 \times 10^{-10}\ N$

Gravitational Fields

Gravity is a non-contact, or field, force. Its effects are observed without the two objects coming into contact with each other. Exactly how this happens is a mystery to this day, but scientists have come up with a mental construct known as a vector field to help understand how gravity and other field forces work.

A vector field describes the value of some physical vector quantity at a specific point in time in space. Envision an object with a gravitational field, such as the planet Earth. The closer other masses are to Earth, the more gravitational force they will experience. You can characterize this by calculating the amount of force the Earth will exert per unit mass at various distances from the Earth. For example, right now, at the location where you are sitting on the surface of the Earth, the gravitational field strength (**g**) is approximately 9.8 newtons of force toward the center of the Earth for every kilogram of gravitational mass, or 9.8 N/kg. If your gravitational mass is 60kg, then the gravitational force you are experiencing is (60kg)×(9.8 N/kg) or 588 newtons down. More commonly, you refer to this as your weight.

Obviously, the closer an object is to the Earth, the larger a gravitational force it will experience, and the farther it is from the Earth, the smaller a gravitational force it will experience.

Attempting to visualize this, picture the strength of the gravitational force on a test object represented by a vector arrow at the position of the object. The denser the force vectors are, the stronger the force, the stronger the "gravitational field." As these field lines become less dense, the gravitational field gets weaker.

If more than one source of a vector field exists, the net field value at a particular point in time and space can be determined by vector addition of vector fields due to each individual source. Visualize a universe in which only two suns exist. Determining the net gravitational field anywhere in the universe would be as simple as determining the gravitational field due to the first sun at the point in space, determining the gravitational field due to the second sun at the point in space, and then adding up the two gravitational field vectors.

A gravitational field (**g**) acting on an object of mass m produces a gravitational force of attraction mg in the direction of the gravitational field vector. In mathematical terms:

$g = G\frac{m}{r^2}$

$$F_g = m\vec{g} \rightarrow \vec{g} = \frac{\vec{F}_g}{m}$$

Combining this equation with Newton's Law of Universal Gravitation, you can derive a general relationship for the gravitational field strength outside a spherically symmetric object.

$$|\vec{g}| = \frac{|\vec{F}_g|}{m} = \frac{G\frac{m_1 m_2}{r^2}}{m} = G\frac{m}{r^2}$$

In the region near Earth's surface, the gravitational field strength **g** is relatively constant. This has allowed us to approximate the gravitational force, or weight, of objects near the surface of the earth as mg up to this point. This approximation breaks down if the gravitational field strength is no longer constant, such as when an object is positioned far from the surface of the Earth.

But wait, you might say — I thought **g** was the acceleration due to gravity on the surface of the Earth! And you would be right. Not only is **g** the gravitational field strength, it's also the acceleration due to gravity. The units even work out. The units of gravitational field strength, N/kg, are equivalent to the units for acceleration, m/s²!

Still skeptical? Try to calculate the gravitational field strength on the surface of the Earth using the knowledge that the mass of the Earth is approximately 5.98×10^{24} kg and the distance from the surface to the center of mass of the Earth (which varies slightly since the Earth isn't a perfect sphere) is approximately 6378 km.

$$g = \frac{Gm}{r^2} = \frac{(6.67 \times 10^{-11} \, N \bullet m^2/kg^2)(5.98 \times 10^{24} \, kg)}{(6378000m)^2} = 9.8 \, m/s^2$$

As expected, the gravitational field strength on the surface of the Earth is the acceleration due to gravity.

8.6 Q: Suppose a 100-kg astronaut feels a gravitational force of 700N when placed in the gravitational field of a planet.

A) What is the magnitude of the gravitational field strength at the location of the astronaut?

B) What is the mass of the planet if the astronaut is 2×10^6 m from its center?

8.6 A: (A) $g = \dfrac{F_g}{m} = \dfrac{700\,N}{100\,kg} = 7\,{}^{N}\!/_{kg}$

(B) $F_g = \dfrac{G\,m_1\,m_2}{r^2}$ $\qquad m_{planet} = \dfrac{F_g\,r^2}{G\,m_{astronaut}}$

$m_{planet} = \dfrac{(700\,N)(2 \times 10^6\,m)^2}{(6.67 \times 10^{-11}\,{}^{N\bullet m^2}\!/_{kg^2})(100\,kg)} = 4.2 \times 10^{23}\,kg$

8.7 Q: What is the acceleration due to gravity at a location where a 15-kilogram mass weighs 45 newtons?

(A) 675 m/s²

(B) 9.81 m/s²

(C) 3.00 m/s²

(D) 0.333 m/s²

8.7 A: (C) $g = \dfrac{F_g}{m} = \dfrac{45\,N}{15\,kg} = 3\,{}^{N}\!/_{kg} = 3\,{}^{m}\!/_{s^2}$

8.8 Q: A 2.0-kilogram object is falling freely near Earth's surface. What is the magnitude of the gravitational force that Earth exerts on the object?

(A) 20 N

(B) 2.0 N

(C) 0.20 N

(D) 0.0 N

8.8 A: (A) 20 N (and the object exerts 20N of force on the Earth!)

8.9 Q: A 1200-kilogram space vehicle travels at 4.8 meters per second along the level surface of Mars. If the magnitude of the gravitational field strength on the surface of Mars is 3.7 newtons per kilogram, the magnitude of the normal force acting on the vehicle is

(A) 320 N

(B) 930 N

(C) 4400 N

(D) 5800 N

8.9 A: (C) To solve this problem, you must recognize that the gravitational field strength of 3.7 N/kg is equivalent to the acceleration due to gravity on Mars, therefore a=3.7 m/s². Then, because the space vehicle isn't accelerating vertically, the normal force must balance the vehicle's weight.

$$F_N = F_g = mg = (1200kg)(3.7 \, ^m\!/_{s^2}) = 4440\,N$$

8.10 Q: A 2.00-kilogram object weighs 19.6 newtons on Earth. If the acceleration due to gravity on Mars is 3.71 meters per second², what is the object's mass on Mars?

(A) 2.64 kg

(B) 2.00 kg

(C) 19.6 N

(D) 7.42 N

8.10 A: (B) 2.00 kg. Mass does not change.

8.11 Q: The graph below represents the relationship between gravitational force and mass for objects near the surface of Earth.

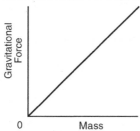

The slope of the graph represents the

(A) acceleration due to gravity

(B) universal gravitational constant

(C) momentum of objects

(D) weight of objects

8.11 A: (A) acceleration due to gravity.

8.12 Q: An alien drops a rock from a height of 2 meters above the surface of the planet Neptune. If the mass of Neptune is 10²⁶ kg and it has a diameter of 49,200 km, how long does it take the rock to fall to the planet's surface?

8.12 A: First, determine the acceleration due to gravity on the surface of Neptune:

$$g = \frac{Gm}{r^2} = \frac{(6.67 \times 10^{-11} \, ^{N \cdot m^2}/_{kg^2})(1 \times 10^{26} \, kg)}{(24,600,000m)^2} = 11\,^{N}/_{kg} = 11\,^{m}/_{s^2}$$

Then, use kinematics to determine the time to fall.

$$y = y_0 + v_{y0}t + \tfrac{1}{2}a_y t^2 \xrightarrow[v_{y0}=0]{y_0=0} y = \tfrac{1}{2}a_y t^2 \rightarrow$$

$$t = \sqrt{\frac{2y}{a_y}} = \sqrt{\frac{2(2m)}{11\,^{m}/_{s^2}}} = 0.6s$$

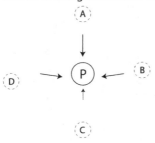

8.13 Q: A gravitational probe in space makes the following incomplete sketch of the gravitational field vectors surrounding planet P.

Based on this map, which of the given locations A, B, C, or D is most likely to contain a massive object?

8.13 A: (C) is most likely to contain a massive object based upon the position of the gravitational field vectors.

Gravitational Potential Energy

Back in Chapter 5 you learned that gravitational potential energy is the energy an object possesses because of its position in a gravitational field, and performed some basic calculations involving constant gravitational fields. You can generalize and expand this relationship by tying in Newton's Law of Universal Gravitation to find the gravitational potential energy existing due to the interaction of any two objects.

Assume two massive objects exist in space. Newton's Law of Universal Gravitation states that the objects will attract each other with a force F_g. As these two objects are attracted to each other, they cause a displacement r until the objects meet. In this way, the gravitational force is doing work on the objects. This work corresponds to the gravitational potential energy of the system in its initial position.

$$\Delta E = F_{\parallel} r = G \frac{m_1 m_2}{r^2} r = \frac{Gm_1 m_2}{r}$$

As this is a relative measure of potential energy, for absolute values to have any physical meaning, a fixed reference point has to be set. Traditionally, a reference point of 0 gravitational potential energy is defined as the gravitational potential energy that exists when any objects are an infinite distance apart from each other. As the objects get closer together, the magnitude of their gravitational potential energy increases in magnitude, but is defined as a negative, indicating that the objects are "caught" in each other's gravitational field. Therefore, absolute gravitational potential energies are negative by convention, and are written as:

$$U_g = -\frac{Gm_1 m_2}{r}$$

8.14 Q: A binary star is a celestial phenomenon in which two stars orbit around their mutual center of mass. If the mass of the primary star is 4M, and the mass of the secondary star is M, what is the gravitational potential energy of the binary star system if the stars are separated by distance r?

8.14 A: $U_g = -\dfrac{Gm_1 m_2}{r} = -\dfrac{G(4M)(M)}{r} = -\dfrac{4GM^2}{r}$

8.15 Q: A rocket is launched vertically from the surface of the Earth with an initial velocity of 10 km/s. What maximum height does the rocket reach, neglecting air resistance? Note that the mass of the Earth is 6×10^{24} kg and the radius of the Earth is 6.37×10^6 m. You may not assume that the acceleration due to gravity is a constant.

8.15 A: Using a conservation of energy approach, the initial total mechanical energy at the time of the rocket's launch must equal the rocket's total mechanical energy when it reaches its highest point. Therefore you can write:

$$K_i + U_i = K_f + U_f$$

The problem then becomes an exercise in algebra, solving for R at the highest point of the rocket's flight and recognizing that the rocket has a kinetic energy of 0 at its highest point. Call the mass of the Earth m_1, the mass of the rocket m_2, the radius of the Earth R_E, and the radius of the rocket R.

$$\tfrac{1}{2}m_2 v^2 + \frac{-Gm_1 m_2}{R_E} = 0 + \frac{-Gm_1 m_2}{R} \rightarrow \frac{1}{R} = \frac{-m_2 v^2}{2Gm_1 m_2} + \frac{Gm_1 m_2}{Gm_1 m_2 R_E} \rightarrow$$

$$\frac{1}{R} = \frac{-v^2}{2Gm_1} + \frac{1}{R_E} \rightarrow \frac{1}{R} = \frac{-(10,000\,^m/_s)^2}{2(6.67\times10^{-11}\,^{N\bullet m^2}/_{kg^2})(6\times10^{24}\,kg)} + \frac{1}{6.37\times10^6\,m} \rightarrow$$

$$R = 3.1\times10^7\,m$$

Having now found R, the maximum distance the rocket reaches from the center of the Earth, you can solve for h, the height of the rocket above Earth's surface.

$$h = R - R_E = 3.1\times10^7\,m - 6.37\times10^6\,m = 2.5\times10^7\,m$$

Orbits

How do celestial bodies orbit each other? The moon orbits the Earth. The Earth orbits the sun. Earth's solar system is in orbit in the Milky Way galaxy... but how does it all work?

To explain orbits, Sir Isaac Newton developed a "thought experiment" in which he imagined a cannon placed on top of a very tall mountain, so tall, in fact, that the peak of the mountain was above the atmosphere (this is important because it allows us to neglect air resistance). If the cannon then launched a projectile horizontally, the projectile would follow a parabolic path to the surface of the Earth.

If the projectile was launched with a higher speed, however, it would travel farther across the surface of the Earth before reaching the ground. If its speed could be increased high enough, the projectile would fall at the same rate the Earth's surface curves away. The projectile would continue falling forever as it circled the Earth! This circular motion describes an orbit.

Put another way, the astronauts in the space shuttle aren't weightless. Far from it, actually; the Earth's gravity is still acting on them and pulling them toward the center of the Earth with a substantial force. You can even calculate that force and the acceleration due to gravity while they're in orbit.

8.16 Q: If the space shuttle orbits the Earth at an altitude of 380 km, what is the gravitational field strength due to the Earth at that altitude? The mass of the Earth is 5.98×10^{24} kg, and the radius of the Earth is approximately 6.37×10^6 m.

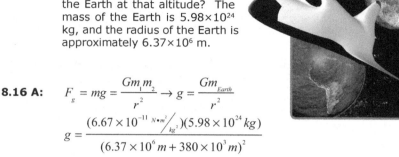

8.16 A:
$$F_g = mg = \frac{Gm_1 m_2}{r^2} \rightarrow g = \frac{Gm_{Earth}}{r^2}$$

$$g = \frac{(6.67 \times 10^{-11} \ ^{N \bullet m^2}/_{kg^2})(5.98 \times 10^{24} \ kg)}{(6.37 \times 10^6 \ m + 380 \times 10^3 \ m)^2}$$

$$g = 8.75 \ ^N/_{kg} = 8.75 \ ^m/_{s^2}$$

This means that the acceleration due to gravity at the altitude the astronauts are orbiting the earth is only 11% less than on the surface of the Earth! In actuality, the space shuttle is falling, but it's moving so fast horizontally that by the time it falls, the Earth has curved away underneath it so that the shuttle remains at the same distance from the center of the Earth. It is in orbit! Of course, this takes tremendous speeds. To maintain an orbit of 380 km, the space shuttle travels approximately 7680 m/s, more than 23 times the speed of sound at sea level!

An object orbiting a celestial body such as a planet in a circular orbit must be held in its circular path by a centripetal force. The centripetal force causing this circular motion is, of course, gravity. You can use this knowledge to determine the speed of any satellite in a circular orbit as a function of the mass of the object it is orbiting and the radius of its orbit.

8.17 Q: A satellite with mass m_2 travels with velocity v in a circular orbit of radius R around an object with mass m_1 as shown in the diagram below.

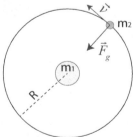

Find the velocity of the orbiting satellite.

8.17 A: First write a Newton's 2nd Law Equation for the centripetal direction.

$$F_{net_c} = F_g = m_2 a_c$$

Next, use Newton's Law of Universal Gravitation to expand the F_g term and replace centripetal acceleration with v^2/R. Then solve for v.

$$F_g = m_2 a_c \xrightarrow[a_c = \frac{v^2}{R}]{F_g = G\frac{m_1 m_2}{R^2}} G\frac{m_1 m_2}{R^2} = m_2 \frac{v^2}{R} \rightarrow$$

$$G\frac{m_1}{R} = v^2 \rightarrow v = \sqrt{\frac{Gm_1}{R}}$$

Note that the velocity of the satellite is independent of the satellite's mass!

Geosynchronous orbit occurs when a satellite maintains an orbit which has a period of one day, so that at the same time each day, the satellite is above the same point on the Earth. **Geostationary orbits** are orbits directly above the equator, which allow satellites to remain above the same point on the Earth (the satellite's rotational velocity matches the Earth's rotational velocity). These are very useful for communications satellites. To "speed up" or "slow down" an orbiting satellite relative to the surface of the Earth, you just change the satellite's altitude.

8.18 Q: Given the mass of the Earth is approximately 6×10^{24} kg, and the radius of the Earth is approximately 6.37×10^6 m, determine the speed and the height above the surface of the Earth for a satellite in a geostationary orbit.

8.18 A: For an object in geostationary orbit, the period of its orbit must be 24 hours, or 86,400 seconds. Using this, you can develop an equation for the speed of the satellite as a function of R.

$$v = \frac{d}{t} = \frac{2\pi R}{T}$$

Next, utilize the relationship you developed from the previous problem between the velocity of a satellite and its orbital radius, and solve for the velocity of the satellite.

$$v = \sqrt{\frac{Gm_1}{R}} \xrightarrow[R = \frac{vT}{2\pi}]{v = \frac{2\pi R}{T}} v = \sqrt{\frac{2\pi Gm_1}{vT}} \rightarrow v = \sqrt[3]{\frac{2\pi Gm_1}{T}} \rightarrow$$

$$v = \sqrt[3]{\frac{2\pi(6.67 \times 10^{-11} \, N \bullet m^2/kg^2)(6 \times 10^{24} \, kg)}{86,400s}} = 3080 \, m/s$$

Then, solve for the radius of the satellite.

$$v = \frac{2\pi R}{T} \rightarrow R = \frac{vT}{2\pi} = \frac{(3080 \, m/s)(86,400s)}{2\pi} = 4.2 \times 10^7 \, m$$

Finally, solve for the height above the surface of the Earth by subtracting the radius of the Earth from the radius of the satellite's orbit.

$$h = R - R_E = 4.2 \times 10^7 \, m - 6.37 \times 10^6 \, m = 3.6 \times 10^7 \, m$$

Escape velocity is the velocity required for an object to completely escape the influence of gravity, which theoretically occurs at an infinite distance from all other masses. You can find the escape velocity of an object by recognizing that the gravitational potential energy of the system is zero when an object has completely escaped gravity's influence, and it has no leftover kinetic energy (its speed is zero).

8.19 Q: Determine the escape velocity for an object of mass m_2 initially held at a distance R from the center of mass of a large celestial body of mass m_1.

8.19 A: Begin by recognizing that the total energy of the system is zero when there is no gravitational potential energy (no gravitational influence) and the object's kinetic energy is zero (speed is zero). Use this relationship to solve for the initial velocity required to reach this state.

$$E = K + U = 0 \rightarrow \tfrac{1}{2}m_2v^2 - \frac{Gm_1m_2}{R} = 0 \rightarrow \tfrac{1}{2}m_2v^2 = \frac{Gm_1m_2}{R} \rightarrow$$

$$v^2 = \frac{2Gm_1m_2}{m_2R} = \frac{2Gm_1}{R} \rightarrow v_{escape} = \sqrt{\frac{2Gm_1}{R}}$$

8.20 Q: The diagram below represents two satellites of equal mass, A and B, in circular orbits around a planet.

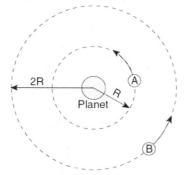

Compared to the magnitude of the gravitational force of attraction between satellite A and the planet, the magnitude of the gravitational force of attraction between satellite B and the planet is

(A) half as great

(B) twice as great

(C) one-fourth as great

(D) four times as great

8.20 A: (C) one-fourth as great due to the inverse square law relationship.

8.21 Q: Calculate the magnitude of the centripetal force acting on Earth as it orbits the Sun, assuming a circular orbit and an orbital speed of 3.00×10⁴ meters per second.

8.21 A: Look up the mass of the Earth and the radius of the Earth's orbit.

$$F_c = ma_c = m\frac{v^2}{r}$$

$$F_c = (5.98 \times 10^{24}\, kg)\frac{(3\times10^4\, ^m\!/_s)^2}{(1.5\times10^{11}\, m)} = 3.59 \times 10^{22}\, N$$

8.22 Q: Determine the total mechanical energy for a satellite of mass m_2 in orbit around a much larger object of mass m_1 in terms of the two masses and the distance between their centers of mass.

8.22 A: Begin by recognizing that the total mechanical energy is the sum of the kinetic and potential energies.

$E = K + U$

Next, substitute in expressions for the kinetic and gravitational potential energies.

$E = \tfrac{1}{2}m_2 v^2 - \dfrac{Gm_1 m_2}{R}$

Finally, utilize the relationship for the velocity of a satellite in orbit to remove the equation's dependence on v and solve for the total energy.

$$E = \tfrac{1}{2}m_2 v^2 - \frac{Gm_1 m_2}{R} \xrightarrow[\substack{v=\sqrt{\frac{Gm_1}{R}} \\ v^2 = \frac{Gm_1}{R}}]{} E = \tfrac{1}{2}m_2\frac{Gm_1}{R} - \frac{Gm_1 m_2}{R} \rightarrow$$

$$E = -\frac{Gm_1 m_2}{2R}$$

Kepler's Laws of Planetary Motion*

In the early 1600s, most of the scientific world believed that the planets should have circular orbits, and many believed that the Earth was the center of the solar system. Using data collected by Tycho Brahe, German astronomer Johannes Kepler developed three laws governing the motion of planetary bodies, which described their orbits as ellipses with the sun at one of the focal points (even though the orbits of many planets are nearly circular). These laws are known as **Kepler's Laws of Planetary Motion**.

Kepler's First Law of Planetary Motion states that the orbits of planetary bodies are ellipses with the sun at one of the two foci of the ellipse.

Kepler's Second Law of Planetary Motion states that if you were to draw a line from the sun to the orbiting body, the body would sweep out equal areas along the ellipse in equal amounts of time. This is easier to observe graphically. In the diagram, if the orbiting body moves from point 1 to point 2 in the same amount of time as it moves from point 3 to point 4, then areas A1 and A2 must also be equal.

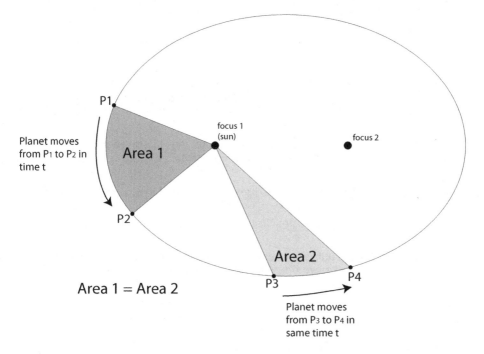

8.23 Q: Given the elliptical planetary orbit shown above, identify the interval during which the planet travels with the highest speed.

(A) Interval P1 to P2

(B) Interval P3 to P4

(C) They are the same.

8.23 A: (A) Interval P1 to P2. Because Area 1 is equal to Area 2, we know that the time interval from P1 to P2 must be equal to the time interval from P3 to P4 by Kepler's 2nd Law of Planetary Motion. Since the planet travels a greater distance from P1 to P2, it must have the higher speed during this portion of its journey.

Note that Kepler's 2nd Law of Planetary Motion is really an application of the law of conservation of angular momentum. If you analyze the system from the perspective of reference point P as shown in the diagram below, and there are no external forces or torques (the only force is gravity through point P), then angular momentum must be conserved.

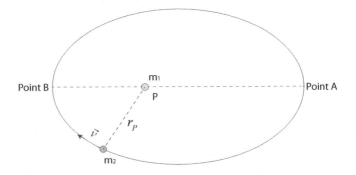

You can then write a conservation of angular momentum equation to obtain a relationship for velocity and radius for any point in orbit:

$$\left|L_P\right| = mvr\sin\theta \rightarrow m_2 v_A r_A \sin\theta_A = m_2 v_B r_B \sin\theta_B$$

At points A and B, however, known as apogee and perigee, respectively, θ=90°, so the analysis for these specific points simplifies to:

$$v_A r_A = v_B r_B$$

Consider mass m_1 as the sun and mass m_2 as an orbiting planet. The planet is farther from the sun at point A (apogee), therefore it moves slower than when it is at point B (perigee).

8.24 Q: Which planet takes the longest amount of time to make one complete revolution around the sun?

(A) Venus

(B) Earth

(C) Jupiter

(D) Uranus

8.24 A: (D) Uranus must have the longest orbital period since it is farthest from the sun according to Kepler's 3rd Law of Planetary Motion.

Kepler's 3rd Law of Planetary Motion, described several years after the first two laws were published, states that the ratio of the squares of the periods of two planets is equal to the ratio of the cubes of their orbital radii. If this sounds confusing, don't worry! Once again it's not as bad as it looks, and is something that is quite easy to derive. In examining Kepler's 3rd Law, we'll have to make an assumption that planets move in circular orbits. This is a reasonable approximation for the planets closest to our sun.

The period (T) is the time it takes for a satellite to make one complete revolution, or travel one circumference, of its circular path. You can solve for the period using the velocity for a satellite in orbit that you found earlier in this chapter.

$$2\pi R = vT \rightarrow T = \frac{2\pi R}{v} \xrightarrow{v=\sqrt{\frac{Gm_1}{R}}} T = \frac{2\pi R\sqrt{R}}{\sqrt{Gm_1}}$$

Next, you can clean up the expression by squaring both sides and dividing by R^3.

$$T = \frac{2\pi R\sqrt{R}}{\sqrt{Gm_1}} \rightarrow T^2 = \frac{4\pi^2 R^3}{Gm_1} \rightarrow \frac{T^2}{R^3} = \frac{4\pi^2}{Gm_1}$$

This ratio, T^2/R^3, is a constant for any orbit! In our solar system, for example, T^2/R^3 is approximately equal to 2.97×10^{-19} s²/m³.

What this is really saying, then, is that planets that are closer to the sun (with a smaller orbital radius) have much shorter periods than planets that are farther from the sun. For example, the planet Mercury, closest to the sun, has an orbital period of 88 days. Neptune, which is 30 times farther from the sun than Earth, has an orbital period of 165 Earth years.

8.25 Q: A satellite orbits a planet in an elliptical path as shown. Specific positions of the satellite are noted on the diagram as A, B, C, and D.

Rank from highest to lowest the following characteristics of the satellite at each position.

I) Speed

II) Gravitational Potential Energy

III) Total Mechanical Energy

Chapter 8: Gravity

8.25 A: I) Speed: C, D, A, B (Kepler's 2nd / 3rd Laws)

II) U_g: B, A, D, C (law of conservation of energy)

III) E_T: same for all (law of conservation of energy)

8.26 Q: Given the orbital radius of Mercury is roughly 5.8×10^{10} m, estimate the period of Mercury's orbit in terms of Earth years.

8.26 A: First, utilize Kepler's 3rd Law to find the ratio of T^2/R^3.

$$\frac{T^2}{R^3} = \frac{4\pi^2}{Gm_1} = 2.97 \times 10^{-19} \, {}^{s^2}\!/_{m^3}$$

Next, solve for the period of Mercury's orbit in seconds.

$$T^2 = R^3 \times 2.97 \times 10^{-19} \, {}^{s^2}\!/_{m^3} = (5.8 \times 10^{10} \, m)^3 \times 2.97 \times 10^{-19} \, {}^{s^2}\!/_{m^3} \rightarrow$$

$$T = 7.6 \times 10^6 \, s$$

Finally, convert Mercury's period in seconds into Earth years.

$$7.6 \times 10^6 \, s \times \frac{1 \, yr}{3.16 \times 10^7 \, s} = 0.24 \, yr$$

8.27 Q: The shape of Mars' orbit around the sun is most accurately described as a:

(A) circle

(B) ellipse

(C) parabola

(D) hyperbola

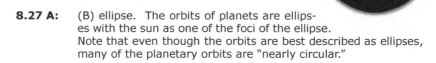

8.27 A: (B) ellipse. The orbits of planets are ellipses with the sun as one of the foci of the ellipse. Note that even though the orbits are best described as ellipses, many of the planetary orbits are "nearly circular."

Test Your Understanding

1. Design an experiment in which an object's inertial mass is measured. Design a separate experiment in which the object's gravitational mass is measured. How would you expect the results of each experiment to compare? How might you account for any differences?

2. Determine the gravitational force of attraction between you and the nearest tangible living organism (yes, lab partners count).

3. How far above the Earth's surface would you have to travel for your weight to be one-half its value on the surface of the Earth?

4. How far above the Earth's surface would you have to travel for your mass to be one-half its value on the surface of the Earth?

5. Determine the minimum initial velocity required for a rocket on the surface of the Earth to completely escape Earth's gravitational influence.

6. Why doesn't the moon crash into the Earth?

7. Explain how Kepler's 2nd Law of Planetary Motion is actually a statement of the law of conservation of angular momentum.

Chapter 9: Oscillations

"Now if this electron is displaced from its equilibrium position, a force that is directly proportional to the displacement restores it like a pendulum to its position of rest."

— *Pieter Zeeman*

Objectives

1. Describe the conditions necessary for simple harmonic motion.
2. Write down an appropriate expression for displacement of the form Acos(ωt) or Asin(ωt) describing the motion.
3. Apply the relationship between frequency, angular frequency, and period.
4. Determine the total energy of an object undergoing simple harmonic motion, and sketch graphs of kinetic and potential energies as functions of time or displacement.
5. Identify the minima, maxima, and zeros of displacement, velocity, and acceleration for an object undergoing simple harmonic motion.
6. Analyze the simple harmonic motion of a spring-block oscillator.
7. Analyze the simple harmonic motion of an ideal pendulum.

Simple Harmonic Motion

When an object is displaced, it may be subject to a restoring force, resulting in a periodic oscillating motion. If the displacement is directly proportional to the linear restoring force, the object undergoes **simple harmonic motion** (SHM). Examples of linear restoring forces causing simple harmonic motion include a mass on a spring, the pendulum on a grandfather clock, a tree limb oscillating after you brush it walking through the woods, a child on a swing, even the vibrations of atoms in a solid can be modeled as simple harmonic motion.

Uniform circular motion also has ties to simple harmonic motion. Consider an object moving in a horizontal circle of radius A at constant angular velocity ω as shown in the diagram below. At any given point in time, the x-position of the object can be described by x=Acosθ, and the y-position can be described by y=Asinθ.

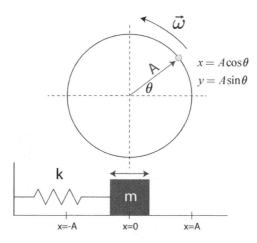

Now imagine a block on a frictionless surface attached to a wall by a spring. If that block is displaced from its equilibrium position by an amount A, and released, its simple harmonic motion along the frictionless surface will mirror the x-motion of the object moving in uniform circular motion. Simple harmonic motion and uniform circular motion are very closely related, as simple harmonic motion describes one dimension of an object moving in uniform circular motion!

Referring back to the chapter on rotational motion, recall that the angular displacement of an object θ is equal to the angular velocity multiplied by the time interval. You can therefore write the x-coordinate of the objects in UCM and SHM as

$$x = A\cos\theta \xrightarrow{\theta=\omega t} x = A\cos(\omega t)$$

In this equation, ω is known as the **angular frequency**, corresponding to the number of radians per second for an object traveling in uniform circular motion. You can convert from angular frequency to frequency and period using the equation:

$$\omega = 2\pi f = \frac{2\pi}{T}$$

Therefore a general form describing the motion of an object undergoing SHM can be written as:

$$x(t) = A\cos(\omega t + \phi)$$

The symbol phi (Φ) refers to the **phase angle**, or starting point, of a sine or cosine curve. Use the cosine curve if the motion starts at maximum amplitude. Use the sine curve if the motion starts at x=0. If the motion being described starts somewhere between maximum amplitude and equilibrium (x=0), you'll have to add in a phase angle component.

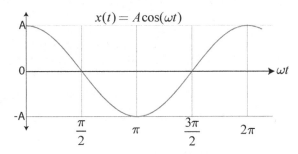

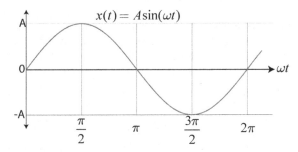

If the maximum displacement of an object undergoing SHM is A, then the maximum speed of the object is ωA, which you can find by taking the slope of the position-time curve, and its maximum acceleration is ω²A, which you can find by taking the slope of the velocity-time curve.

9.01 Q: An oscillating system is created by a releasing an object from a maximum displacement of 0.2 meters. The object makes 60 complete oscillations in one minute. Determine the object's angular frequency.

9.01 A: $\omega = 2\pi f = 2\pi(1Hz) = 2\pi\,{}^{rad}\!/_{s}$

9.02 Q: Referring to problem 9.01, determine the object's position at time t=10 seconds.

9.02 A: $x = A\cos(\omega t)\xrightarrow[A=0.2\,m]{\omega=2\pi^{rad}/_s} x = (0.2m)\cos(2\pi t)\xrightarrow{t=10s}$

$x = (0.2m)\cos(2\pi\,^{rad}/_s \times 10s) = 0.2m$

9.03 Q: Referring to problem 9.01, determine the time when the object is at position x=0.1m.

9.03 A: $x = A\cos(2\pi t) \rightarrow \cos(2\pi t) = \dfrac{x}{A} \rightarrow 2\pi t = \cos^{-1}\left(\dfrac{x}{A}\right) \rightarrow$

$t = \dfrac{\cos^{-1}\left(\dfrac{x}{A}\right)}{2\pi} = \dfrac{\cos^{-1}\left(\dfrac{0.1m}{0.2m}\right)}{2\pi\,^{rad}/_s} = 0.167s$

Note: if you arrived at an answer of roughly 9.55 seconds, make sure your calculator is functioning in "Radian" mode instead of "Degree" mode for these trigonometric functions.

Horizontal Spring-Block Oscillators

A popular demonstration vehicle for simple harmonic motion is the spring-block oscillator. The horizontal spring-block oscillator consists of a block of mass m sitting on a frictionless surface, attached to a vertical wall by a spring of spring constant k, as shown in the diagram below.

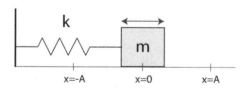

The block is then displaced an amount A from its equilibrium position and allowed to oscillate back and forth. As the block sits on a frictionless surface, in the ideal scenario the block would continue its periodic motion indefinitely.

The angular frequency of the block's oscillation can be determined from the spring constant and the mass.

$$\omega = \sqrt{\frac{k}{m}}$$

Knowing the angular frequency, it's quite straightforward then to calculate the period of oscillation for the spring-block oscillator.

$$T_S = \frac{2\pi}{\omega} \xrightarrow{\omega = \sqrt{\frac{k}{m}}} T_S = 2\pi\sqrt{\frac{m}{k}}$$

Note that the period of oscillation depends only on the mass of the block and the spring constant. There is no dependency on the magnitude of the displacement of the block.

9.04 Q: A 5-kg block is attached to a 2000 N/m spring as shown and displaced a distance of 8 cm from its equilibrium position before being released.

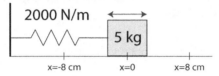

Determine the period of oscillation, the frequency, and the angular frequency for the block.

9.04 A: $T_S = 2\pi\sqrt{\frac{m}{k}} = 2\pi\sqrt{\frac{5kg}{2000\,{}^{N}\!/_{m}}} = 0.314s$

$f = \frac{1}{T} = \frac{1}{0.314s} = 3.18Hz$

$\omega = 2\pi f = 2\pi(3.18Hz) = 20\,{}^{rad}\!/_{s}$

9.05 Q: Rank the following horizontal spring-block oscillators resting on frictionless surfaces in terms of their period, from longest to shortest.

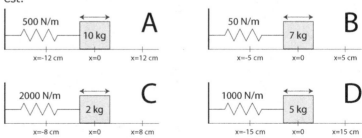

9.05 A: B, A, D, C

It's also interesting to look at the energy of the spring-block oscillator while its undergoing simple harmonic motion. Because the surface is frictionless, the total energy of the system remains constant. However, there is a continual transfer of kinetic energy into elastic potential energy and back.

When the block is at its equilibrium position, there is no elastic potential energy stored in the spring, therefore all of the energy of the block is kinetic. The block has achieved its maximum speed. At this position, there is also no net force on the block, therefore the block's acceleration is zero.

When the block is at its maximum amplitude position, all of its energy is stored in the spring as elastic potential energy. For an instant its kinetic energy is zero, therefore its velocity is zero. Further, at this position, the spring exhibits a maximum force on the block, providing the maximum acceleration. Let's take a look at this graphically by examining a spring-block oscillator at various points in its periodic path.

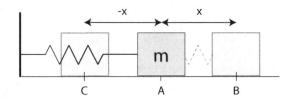

Graphs of Related Physical Quantities

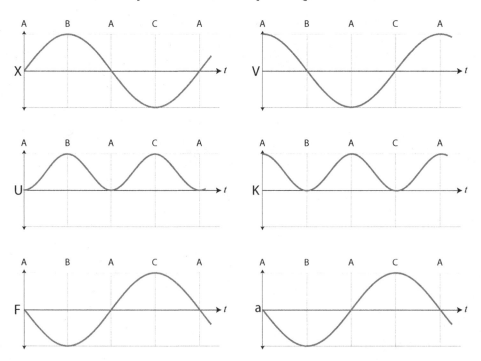

	A	B	C
Displacement (x)	0	X	-X
Velocity (v)	max	0	0
Potential Energy (U)	0	max	max
Kinetic Energy (K)	max	0	0
Force (F)	0	-max	max
Acceleration (a)	0	-max	max

Of course, through the entire time interval, the total mechanical energy of the spring-block oscillator remains constant.

In reality, some energy is typically lost to friction or other non-conservative forces. Over time, the amplitude of the oscillations steadily decrease. This is known as **damping**, or damped harmonic motion.

9.06 Q: A 2-kg block is attached to a spring. A force of 20 newtons stretches the spring to a displacement of 50 cm. Find:

A) the spring constant

B) the total energy

C) the speed of the block at the equilibrium position

D) the speed of the block at x=30 cm

E) the speed of the block at x=-40 cm

F) the acceleration at the equilibrium position

G) the magnitude of the acceleration at at x=50 cm

H) the net force on the block at the equilibrium position

I) the net force at x=25 cm

J) the position where the kinetic energy is equal to the potential energy.

9.06 A:

A) $k = \dfrac{F}{x} = \dfrac{20N}{0.5m} = 40\,{}^{N}\!/_{m}$

B) $E_T = U_S = \frac{1}{2}kx^2 = \frac{1}{2}(40\,{}^{N}\!/_{m})(0.5m)^2 = 5J$

C) $U_S = K = \frac{1}{2}mv^2 = E_T \rightarrow v = \sqrt{\dfrac{2E_T}{m}} = \sqrt{\dfrac{2(5J)}{2kg}} = 2.2\,{}^{m}\!/_{s}$

D) $E_T = U_S + K = \frac{1}{2}kx^2 + \frac{1}{2}mv^2 \rightarrow 2E_T - kx^2 = mv^2 \rightarrow$

$v = \sqrt{\dfrac{2E_T - kx^2}{m}} = \sqrt{\dfrac{2(5J) - (40\,{}^{N}\!/_{m})(0.3m)^2}{2kg}} = 1.8\,{}^{m}\!/_{s}$

E) $v = \sqrt{\dfrac{2E_T - kx^2}{m}} = \sqrt{\dfrac{2(5J) - (40\,\text{N}/\text{m})(0.4m)^2}{2kg}} = 1.3\,\text{m}/\text{s}$

F) at x=0, F=0, therefore acceleration = 0

G) $F = ma = -kx \rightarrow a = \dfrac{-kx}{m} = \dfrac{(-40\,\text{N}/\text{m})(0.5m)}{2kg} = -10\,\text{m}/\text{s}^2$

H) At equilibrium, there is no spring displacement, so net force is zero.

I) $F = -kx = (-40\,\text{N}/\text{m})(0.25m) = -10N$

J) $K = U_s = \dfrac{E_T}{2} \rightarrow \tfrac{1}{2}kx^2 = \dfrac{E_T}{2} \rightarrow x = \sqrt{\dfrac{E_T}{k}} = \sqrt{\dfrac{5J}{40\,\text{N}/\text{m}}} = 0.35m$

Note that the point where kinetic energy and potential energy are equal is **NOT** halfway between the equilibrium and maximum displacement points.

Spring Combinations*

It is also possible to attach more than one spring to an object. In these cases, analyses can be simplified considerably by treating the combination of springs as a single spring with an equivalent spring constant.

For springs in parallel, calculate an equivalent spring constant for the system by starting with Hooke's Law, recognizing that displacement is the same for both springs.

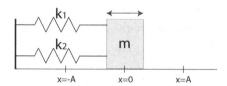

$$F = k_1 x + k_2 x = (k_1 + k_2)x = k_{eq}x \rightarrow k_{eq} = k_1 + k_2$$

For springs in series, you will again calculate an equivalent spring constant for the system, beginning the analysis by realizing the force on each spring must be the same according to Newton's 3rd Law of Motion.

$$F = -k_1 x_1 = -k_2 x_2 \rightarrow x_1 = \frac{k_2}{k_1} x_2$$

Next, recognizing that the total displacement is equal to the sum of the displacement of the springs, you can combine the equations and solve for an equivalent spring constant.

$$F = -k_{eq}(x_1 + x_2) \xrightarrow{\ x_1 = \frac{k_2}{k_1}x_2\ } F = -k_{eq}\left(\frac{k_2}{k_1}x_2 + x_2\right) \xrightarrow{\ F = -k_2 x_2\ }$$

$$k_2 x_2 = k_{eq} x_2 \left(\frac{k_2}{k_1} + 1\right) \rightarrow k_2 = k_{eq}\left(\frac{k_2}{k_1} + 1\right) \rightarrow \frac{1}{k_{eq}} = \frac{1}{k_1} + \frac{1}{k_2}$$

9.07 Q: Rank the spring block oscillators in the diagram below from highest to lowest in terms of:

I) equivalent spring constant

II) period of oscillation

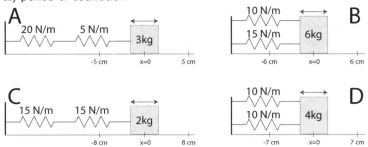

9.07 A: I) B, D, C, A

II) A, C, B, D

Vertical Spring-Block Oscillators

Spring-block oscillators can also be set up vertically as shown in the diagram.

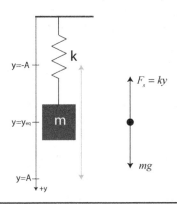

Start your analysis by drawing a Free Body Diagram for the block, noting that gravity pulls the mass down, while the force of the spring provides the upward force. Call down the positive y-direction. At its equilibrium position, $y=y_{eq}$. You can then write a Newton's 2nd Law Equation for the block and solve for y_{eq}.

$$F_{net_y} = mg - ky = ma_y \xrightarrow{equilibrium} mg - ky = 0 \rightarrow y_{eq} = \frac{mg}{k}$$

Once the system has settled at equilibrium, you can displace the mass by pulling it some amount to either +A or lifting it an amount -A. The new system can be analyzed as follows:

$$F_{net_y} = mg - k(y_{eq} + A) = mg - ky_{eq} - kA \xrightarrow{mg - ky_{eq} = 0} F_{net_y} = -kA$$

This is the same analysis you would do for a horizontal spring system with spring constant k displaced an amount A from its equilibrium position. This means, in short, that to analyze a vertical spring system, all you do is find the new equilibrium position of the system, taking into account the effect of gravity, then treat it as a system with only the spring force to deal with, oscillating around the new equilibrium point. No need to continue to deal with the force of gravity!

9.08 Q: A 2-kg block attached to an unstretched spring of spring constant k=200 N/m as shown in the diagram below is released from rest.

I) Determine the period of the block's oscillation.

II) What is the maximum displacement of the block from its equilibrium while undergoing simple harmonic motion?

9.08 A: I) $T_s = 2\pi\sqrt{\frac{m}{k}} = 2\pi\sqrt{\frac{2kg}{200\,{}^N/_m}} = 0.63s$

II) The gravitational potential energy at the block's starting point, $mg\Delta y$, must equal the elastic potential energy stored in the spring at its lowest point. Use this to solve for Δy.

$$U_g = U_s \rightarrow mg\Delta y = \tfrac{1}{2}k\Delta y^2 \rightarrow \Delta y = \frac{2mg}{k}$$

If Δy is the total displacement of the block from its highest point to its lowest point, the maximum displacement of the block from its equilibrium point, A, must be half of Δy.

$$A = \frac{\Delta y}{2} = \frac{mg}{k} = \frac{(2kg)(9.8\,{}^m/_{s^2})}{200\,{}^N/_m} = 0.1m$$

9.09 Q: A 5-kg block is attached to a vertical spring (k=500 N/m). After the block comes to rest, it is pulled down 3 cm and released.

I) What is the period of oscillation?

II) What is the maximum displacement of the spring from its initial unstrained position?

9.09 A: I) $T_S = 2\pi\sqrt{\dfrac{m}{k}} = 2\pi\sqrt{\dfrac{5kg}{500\,^N/_m}} = 0.63s$

II) First determine the displacement of the spring when the block is hanging and at rest.

$$F_{net_y} = 0 \rightarrow kd = mg \rightarrow d = \dfrac{mg}{k} = \dfrac{(5kg)(9.8\,^m/_{s^2})}{500\,^N/_m} = 0.1m$$

Then the block is pulled down 3 cm, so the maximum displacement must be the displacement while the block is at rest in addition to the 3 cm the block is pulled down.

$$y_{max} = 0.1m + 0.03m = 0.13m$$

Ideal Pendulums

Ideal Pendulums provide another demonstration vehicle for simple harmonic motion. Consider a mass m attached to a light string that swings without friction about the vertical equilibrium position. As the mass travels along its path, energy is continuously transferred between gravitational potential energy and kinetic energy. The restoring force in the case of the ideal pendulum is provided by gravity.

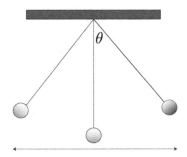

The angular frequency of the ideal pendulum, for small angles of theta, is given by:

$$\omega = \sqrt{\dfrac{g}{l}}$$

You can then find the period of the pendulum for small angles of theta:

$$T_p = \frac{2\pi}{\omega} \xrightarrow{\omega=\sqrt{\frac{g}{l}}} T_p = 2\pi\sqrt{\frac{l}{g}}$$

Notice that the period of the pendulum is dependent only upon the length of the pendulum and the gravitational field strength... there is no mass dependance!

9.10 Q: A grandfather clock is designed such that each swing (or half-period) of the pendulum takes one second. How long is the pendulum in a grandfather clock?

9.10 A: $T_p = 2\pi\sqrt{\dfrac{l}{g}} \rightarrow l = \dfrac{gT^2}{4\pi^2} = \dfrac{(9.8\,{}^{m}\!/_{s^2})(2s)^2}{4\pi^2} = 1m$

9.11 Q: What is the period of a grandfather clock on the moon, where the acceleration due to gravity on the surface is roughly one-sixth that of Earth?

9.11 A: $T_p = 2\pi\sqrt{\dfrac{l}{g}} = 2\pi\sqrt{\dfrac{1m}{\frac{1}{6}(9.8\,{}^{m}\!/_{s^2})}} = 4.9s$

9.12 Q: Rank the following pendulums of uniform mass density from highest to lowest frequency.

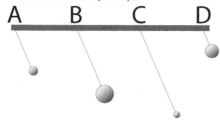

9.12 A: D, A, B, C

Utilizing the geometric representation of the pendulum previously developed in question 5.27, you can construct a diagram detailing the energy and forces acting on the pendulum at various points in its parabolic path.

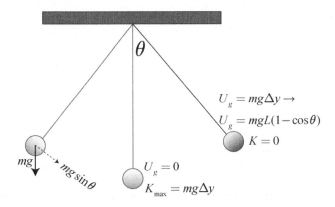

At the highest point, as shown on the left, the mass is being pulled back toward its equilibrium position by gravity. Specifically, the component of gravity along the mass's path, mgsinθ. At the equilibrium position, the gravitational potential energy is at a minimum, and the kinetic energy of the mass is at a maximum. At the highest point, as shown on the right, all the energy is gravitational potential energy again. Using the law of conservation of energy, solving for the maximum velocity of the mass at its lowest position is quite straightforward.

$$K = U_g \rightarrow \tfrac{1}{2}mv^2 = mgL(1-\cos\theta) \rightarrow v = \sqrt{2gL(1-\cos\theta}$$

Further, from this same graph you can create a graph of kinetic energy, gravitational potential energy, and total energy as a position of the mass along the x-axis.

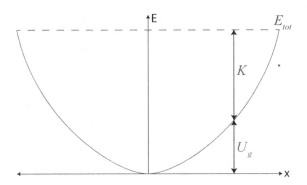

The magnitude of the energy below the parabolic line represents the gravitational potential energy of the mass, while above the line the kinetic energy is represented. The total mechanical energy remains constant throughout the entire path of the pendulum.

9.13 Q: The period of an ideal pendulum is T. If the mass of the pendulum is tripled while its length is quadrupled, what is the new period of the pendulum?

(A) 0.5 T

(B) T

(C) 2T

(D) 4T

9.13 A: (C) 2T

9.14 Q: Which of the following are true for an ideal pendulum consisting of a mass oscillating back and forth on a light string? (Choose all that apply.)

(A) The kinetic energy is always equal to the potential energy.

(B) The maximum force on the pendulum occurs when the pendulum has its maximum kinetic energy.

(C) The tangential acceleration of the pendulum is zero when the mass is at its lowest point.

(D) The angular acceleration of the mass remains constant.

9.14 A: (C) The tangential acceleration of the pendulum is zero when the mass is at its lowest point.

9.15 Q: A pendulum of length 20 cm and mass 1 kg is displaced an angle of 10 degrees from the vertical. What is the maximum speed of the pendulum?

9.15 A: $v = \sqrt{2gL(1-\cos\theta)} = \sqrt{2(9.8\,{}^m\!/_{s^2})(0.2m)(1-\cos 10°)} = 0.24\,{}^m\!/_s$

9.16 Q: A pendulum of length 0.5 m and mass 5 kg is displaced an angle of 14 degrees from the vertical. What is the speed of the pendulum when its angle from the vertical is 7 degrees?

9.16 A: $U_{g_{top}} = U_{g_{bottom}} + K_{bottom} \rightarrow mgL(1-\cos 14°) = mgL(1-\cos 7°) + \frac{1}{2}mv^2 \rightarrow$

$v = \sqrt{2gL(\cos 7° - \cos 14°)} = 0.47\,{}^m\!/_s$

Test Your Understanding

1. Describe at least two methods of determine the spring constant of a spring. Which would you expect to provide more accurate data? Explain.

2. Design an experiment to measure the acceleration due to gravity using a pendulum. What equipment would you require? What data would you collect? What calculations would you make? How would you minimize error?

3. An ideal pendulum is displaced by an angle theta from the vertical. Fill in the following graphs describing the motion of the pendulum in similar fashion as those describing the spring-block oscillator.

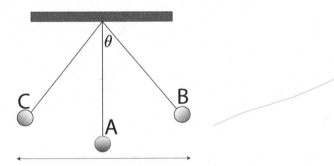

Graphs of Related Physical Quantities

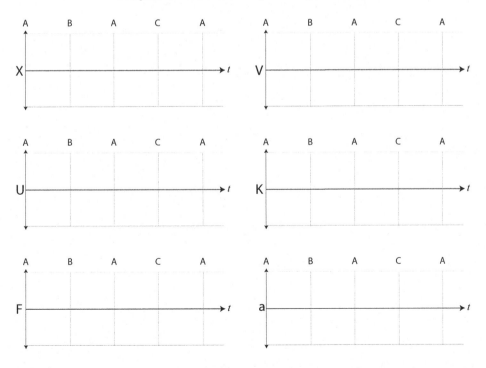

4. A student measures the mass of a block using a spring-block oscillator by measuring the period for the block attached to a variety of springs with known spring constants. What would a graph of t the period vs. spring constant look like? How could you use that graph to determine the mass of the block?

5. How could you double the maximum speed of a 40-kg child on a playground swing?

Chapter 10: Mechanical Waves

*"It would be possible to describe everything scientifically,
but it would make no sense; it would be without meaning,
as if you described a Beethoven symphony
as a variation of wave pressure."*

— *Albert Einstein*

Objectives

1. Define a pulse.
2. Describe the behavior of a pulse at a boundary.
3. Explain the characteristics of transverse and longitudinal waves.
4. Define four terms to describe periodic waves: amplitude, speed, wavelength, and frequency.
5. Understand how the principle of superposition is applied when two pulses meet.
6. Apply the principle of superposition to the phenomenon of interference.
7. Describe the formation of standing waves in strings as well as open and closed tubes.
8. Understand how resonance occurs.
9. Understand the nature of sound waves.
10. Apply the Doppler effect qualitatively to problems involving moving sources or moving observers.
11. Explain the phenomena of beats in terms of interference of waves with slightly differing frequencies.

Waves transfer energy through matter or space, and are found everywhere: sound waves, light waves, microwaves, radio waves, water waves, earthquake waves, slinky waves, x-rays, and on and on. Developing an understanding of waves will allow you to understand how energy is transferred in the universe, and will eventually lead to a better understanding of matter and energy itself!

Wave Characteristics

A **pulse** is a single disturbance which carries energy through a medium or through space. Imagine you and your friend holding opposite ends of a slinky. If you quickly move your arm up and down, a single pulse will travel down the slinky toward your friend.

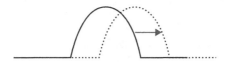

If, instead, you generate several pulses at regular time intervals, you now have a wave carrying energy down the Slinky. A **wave**, therefore is a repeated disturbance which carries energy. The mass of the Slinky doesn't move from one end of the slinky to the other, but the energy it carries does.

When a pulse or wave reaches a hard boundary, it reflects off the boundary, and is inverted. If a pulse or wave reaches a soft, or flexible, boundary, it still reflects off the boundary, but does not invert.

Waves can be classified in several different ways. One type of wave, known as a **mechanical wave**, requires a medium, or material, through which to travel. Examples of mechanical waves include water waves, sound waves, slinky waves, and even seismic waves. **Electromagnetic waves**, on the other hand, do not require a medium in order to travel. Electromagnetic waves (or EM waves) are considered part of the Electromagnetic Spectrum. Examples of EM waves include light, radio waves, microwaves, and even X-rays.

Further, waves can be classified based upon their direction of vibration. Waves in which the "particles" of the wave vibrate in the same direction as the wave velocity are known as **longitudinal**, or compressional, waves. Examples of longitudinal waves include sound waves and seismic P waves. Waves in which the particles of the wave vibrate perpendicular to the wave's direction of motion are known as **transverse** waves. Examples of transverse waves include seismic S waves, electromagnetic waves, and even stadium waves (the "human" waves you see at baseball and football games!).

Video animations of waves reflecting off boundaries as well as longitudinal and transverse waves can be viewed at http://bit.ly/gC1TMU.

Waves have a number of characteristics which define their behavior. Looking at a transverse wave, you can identify specific locations on the wave. The highest points on the wave are known as **crests**. The lowest points on the wave are known as **troughs**. The **amplitude** of the wave, corresponding to the energy of the wave, is the distance from the baseline to a crest or the baseline to a trough.

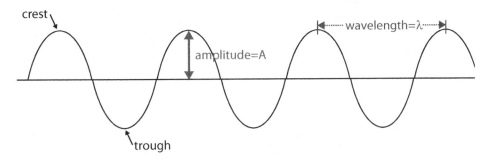

The length of the wave, or **wavelength**, represented by the Greek letter lambda (λ), is the distance between corresponding points on consecutive waves (i.e. crest to crest or trough to trough). Points on the same wave with the same displacement from equilibrium moving in the same direction (such as a crest to a crest or a trough to a trough) are said to be in phase (phase difference is 0° or 360°). Points with opposite displacements from equilibrium (such as a crest to a trough) are said to be 180° out of phase.

10.01 Q: Which type of wave requires a material medium through which to travel?

(A) sound

(B) television

(C) radio

(D) x ray

10.01 A: (A) sound is a mechanical wave and therefore requires a medium.

10.02 Q: The diagram below represents a transverse wave traveling to the right through a medium. Point A represents a particle of the medium.

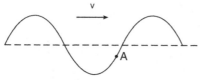

In which direction will particle A move in the next instant of time?

(A) up

(B) down

(C) left

(D) right

10.02 A: (B) particle A will move down as the wave passes.

10.03 Q: As a transverse wave travels through a medium, the individual particles of the medium move

(A) perpendicular to the direction of wave travel

(B) parallel to the direction of wave travel

(C) in circles

(D) in ellipses

10.03 A: (A) perpendicular to the direction of wave travel.

10.04 Q: A ringing bell is located in a chamber. When the air is removed from the chamber, why can the bell be seen vibrating but not be heard?

(A) Light waves can travel through a vacuum, but sound waves cannot.

(B) Sound waves have greater amplitude than light waves.

(C) Light waves travel slower than sound waves.

(D) Sound waves have higher frequency than light waves.

10.04 A: (A) Light is an EM wave, while sound is a mechanical wave.

10.05 Q: A single vibratory disturbance moving through a medium is called
(A) a node
(B) an antinode
(C) a standing wave
(D) a pulse

10.05 A: (D) a pulse.

10.06 Q: A periodic wave transfers
(A) energy, only
(B) mass, only
(C) both energy and mass
(D) neither energy nor mass

10.06 A: (A) energy, only.

10.07 Q: The diagram below represents a transverse wave.

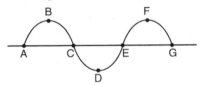

The wavelength of the wave is equal to the distance between points
(A) A and G
(B) B and F
(C) C and E
(D) D and F

10.07 A: (B) B and F is the wavelength as measured from crest to crest.

10.08 Q: The diagram below represents a periodic wave.

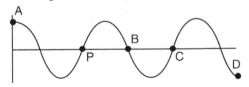

Which point on the wave is in phase with point P?
(A) A
(B) B
(C) C
(D) D

10.08 A: (C) Point C is the same point as point P but on a consecutive wave. Point B doesn't qualify because the wave has not completed a complete cycle.

10.09 Q: The diagram below represents a transverse wave moving on a uniform rope with point A labeled. On the diagram, mark an X at the point on the wave that is 180° out of phase with point A.

10.09 A:

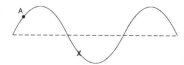

10.10 Q: The diagram below represents a transverse wave.

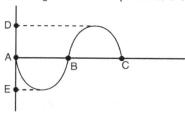

The distance between which two points identifies the amplitude of the wave?

(A) A and B

(B) A and C

(C) A and E

(D) D and E

10.10 A: (C) Amplitude is measured from the baseline to a crest or a trough, therefore amplitude is the distance between A and E.

10.11 Q: The diagram below represents a transverse wave traveling in a string.

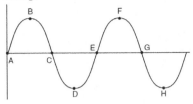

Which two labeled points are 180° out of phase?

(A) A and D

(B) B and F

(C) D and F

(D) D and H

10.11 A: (C) D and F.

In similar fashion, longitudinal waves also have amplitude and wavelength. In the case of longitudinal waves, however, instead of crests and troughs, the longitudinal waves have areas of high density (**compressions**) and areas of low density (**rarefactions**), as shown in the representation of the particles of a sound wave. The wavelength, then, of a compressional wave is the distance between compressions, or the distance between rarefactions. Once again, the amplitude corresponds to the energy of the wave.

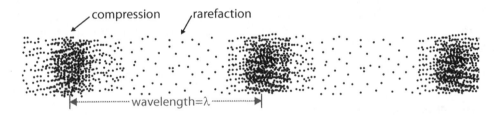

compression   rarefaction

wavelength=λ

10.12 Q: A periodic wave is produced by a vibrating tuning fork. The amplitude of the wave would be greater if the tuning fork were

(A) struck more softly

(B) struck harder

(C) replaced by a lower frequency tuning fork

(D) replaced by a higher frequency tuning fork

10.12 A: (B) Striking the tuning fork harder gives the tuning fork more energy, increasing the sound wave's amplitude.

10.13 Q: Increasing the amplitude of a sound wave produces a sound with
  (A) lower speed
  (B) higher pitch
  (C) shorter wavelength
  (D) greater loudness

10.13 A: (D) greater loudness due to the greater energy / amplitude of the wave.

10.14 Q: A longitudinal wave moves to the right through a uniform medium, as shown below.

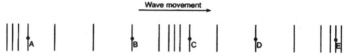

I) Points A, B, C, D, and E represent the positions of particles of the medium. What is the direction of the motion of the particles at position C as the wave moves to the right?

II) Between which two points on the wave could you measure a complete wavelength?

10.14 A: I) The particles move to the left and right at position C, as the particles in a longitudinal wave vibrate parallel to the wave velocity.

II) You could measure a complete wavelength between points A and C, since A and C represent the same point on successive waves.

The Wave Equation

The **frequency** (f) of a wave describes the number of waves that pass a given point in a time period of one second. The higher the frequency, the more waves that pass. Frequency is measured in number of waves per second (1/s), also known as a Hertz (Hz). If 60 waves pass a given point in a second, the frequency of the wave would be 60 Hz.

Closely related to frequency, the **period** (T) of a wave describes how long it takes for a single wave to pass a given point and can be found as the reciprocal of the frequency. Period is a measurement of time, and therefore is measured in seconds.

10.15 Q: What is the period of a 60-hertz electromagnetic wave traveling at 3.0×10^8 meters per second?

10.15 A: $T = \dfrac{1}{f} = \dfrac{1}{60 Hz} = 0.0167s$

10.16 Q: Which unit is equivalent to meters per second?
    (A) Hz•s
    (B) Hz•m
    (C) s/Hz
    (D) m/Hz

10.16 A: (B) $\dfrac{m}{s} = Hz \bullet m$

10.17 Q: The product of a wave's frequency and its period is
    (A) one
    (B) its velocity
    (C) its wavelength
    (D) Planck's constant

10.17 A: (A) $f \bullet T = f \bullet \dfrac{1}{f} = 1$

Because waves move through space, they must have velocity. The velocity of a wave is a function of the type of wave and the medium it travels through. Electromagnetic waves moving through a vacuum, for instance, travel at roughly 3×10^8 m/s. This value is so famous and important in physics it is given its own symbol, **c**. When an electromagnetic wave enters a different medium, such as glass, it slows down. If the same wave were to then re-emerge from glass back into a vacuum, it would again travel at c, or 3×10^8 m/s.

The speed of a wave can be easily related to its frequency and wavelength. Speed of a wave is determined by the wave type and the medium it is traveling through. For a given wave speed, as frequency increases, wavelength must decrease, and vice versa. This can be shown mathematically using the wave equation.

$$v = f\lambda$$

10.18 Q: A periodic wave having a frequency of 5 hertz and a speed of 10 meters per second has a wavelength of

(A) 0.50 m

(B) 2.0 m

(C) 5.0 m

(D) 50. m

10.18 A: (B) $v = f\lambda$

$$\lambda = \frac{v}{f} = \frac{10\,{}^{m}\!/\!_{s}}{5\,Hz} = 2m$$

10.19 Q: If the amplitude of a wave is increased, the frequency of the wave will

(A) decrease

(B) increase

(C) remain the same

10.19 A: (C) remain the same.

10.20 Q: An electromagnetic wave traveling through a vacuum has a wavelength of 1.5×10^{-1} meters. What is the period of this electromagnetic wave?

(A) 5.0×10^{-10} s

(B) 1.5×10^{-1} s

(C) 4.5×10^{7} s

(D) 2.0×10^{9} s

10.20 A: (A) $v = f\lambda = \dfrac{\lambda}{T}$

$$T = \frac{\lambda}{v} = \frac{1.5 \times 10^{-1}\,m}{3 \times 10^{8}\,{}^{m}\!/\!_{s}} = 5 \times 10^{-10}\,s$$

10.21 Q: A surfacing blue whale produces water wave crests having an amplitude of 1.2 meters every 0.40 seconds. If the water wave travels at 4.5 meters per second, the wavelength of the wave is

(A) 1.8 m

(B) 2.4 m

(C) 3.0 m

(D) 11 m

10.21 A: (A) $v = f\lambda$

$$\lambda = \frac{v}{f} = vT = (4.5 \, ^m\!/_s)(0.4s) = 1.8m$$

Sound Waves

Sound is a mechanical wave which is observed by detecting vibrations in the inner ear. Typically, you think of sound as traveling through air, therefore the particles vibrating are air molecules, transferring energy and momentum through the air. Sound can travel through other media as well, including water, wood, and even steel.

The particles of a sound wave vibrate in a direction parallel with the direction of the sound wave's velocity, therefore sound is a longitudinal wave. The speed of sound in air at standard temperature and pressure (STP) is 331 m/s. In dry air at room temperature, the speed of sound is approximately 343 m/s.

10.22 Q: At an outdoor physics demonstration, a delay of 0.50 seconds was observed between the time sound waves left a loudspeaker and the time these sound waves reached a student through the air. If the air is at STP, how far was the student from the speaker?

10.22 A: $\bar{v} = \dfrac{d}{t} \rightarrow d = \bar{v}t = (331 \, ^m\!/_s)(0.50s) = 166m$

10.23 Q: The sound wave produced by a trumpet has a frequency of 440 hertz. What is the distance between successive compressions in this sound wave as it travels through air at STP?
   (A) 1.5×10^{-6} m
   (B) 0.75 m
   (C) 1.3 m
   (D) 6.8×10^5 m

10.23 A: (B) $v = f\lambda$

$$\lambda = \frac{v}{f} = \frac{331 \, ^m\!/_s}{440\,Hz} = 0.75m$$

10.24 Q: A stationary research ship uses sonar to send a 1.18×10^3-hertz sound wave down through the ocean water. The reflected sound wave from the flat ocean bottom 324 meters below the ship is detected 0.425s second after it was sent from the ship.

A) Calculate the speed of the sound wave in the ocean water.

B) Calculate the wavelength of the sound wave in the water.

C) Determine the period of the sound wave in the water.

10.24 A: A) $\bar{v} = \dfrac{d}{t} = \dfrac{648m}{0.425s} = 1520\,^m/_s$

B) $v = f\lambda$

$\lambda = \dfrac{v}{f} = \dfrac{1525\,^m/_s}{1180\,Hz} = 1.29m$

C) $T = \dfrac{1}{f} = \dfrac{1}{1180\,Hz} = 8.47 \times 10^{-4}\,s$

When sound waves are observed through hearing, you pick up the amplitude, or energy, of the waves as loudness. The frequency of the wave is perceived as pitch, with higher frequencies observed as a higher pitch. Typically, humans can hear a frequency range of 20Hz to 20,000 Hz, although young observers can often detect frequencies above 20,000 Hz, an ability which declines with age.

Certain devices create strong sound waves at a single specific frequency. If another object, having the same "**natural frequency**," is impacted by these sound waves, it may begin to vibrate at this frequency, producing more sound waves. The phenomenon where one object emitting a sound wave with a specific frequency causes another object with the same natural frequency to vibrate is known as **resonance**. A dramatic demonstration of resonance involves an opera singer breaking a glass by singing a high pitch note. The singer creates a sound wave with a frequency equal to the natural frequency of the glass, causing the glass to vibrate at its natural, or resonant, frequency so energetically that it shatters. A video of this is available at http://aplusphysics.com/l/shatter. The same effect can be observed when you push someone on a swing. By pushing at the resonant frequency, the swing goes higher and higher!

10.25 Q: Sound waves strike a glass and cause it to shatter. This phenomenon illustrates

(A) resonance   (B) refraction

(C) reflection   (D) diffraction

10.25 A: (A) resonance

10.26 Q: A dampened fingertip rubbed around the rim of a crystal glass causes the glass to vibrate and produce a musical note. This effect is due to

(A) resonance

(B) refraction

(C) reflection

(D) rarefaction

10.26 A: (A) resonance

10.27 Q: Resonance occurs when one vibrating object transfers energy to a second object causing it to vibrate. The energy transfer is most efficient when, compared to the first object, the second object has the same natural

(A) frequency

(B) loudness

(C) amplitude

(D) speed

10.27 A: (A) frequency.

10.28 Q: A car traveling at 70 kilometers per hour accelerates to pass another car.

When the car reaches a speed of 90 kilometers per hour the driver hears the glove compartment door start to vibrate. By the time the speed of the car is 100 kilometers per hour, the glove compartment door has stopped vibrating. This vibrating phenomenon is an example of

(A) destructive interference

(B) the Doppler effect

(C) diffraction

(D) resonance

10.28 A: (D) resonance.

10.29 Q: Which wave phenomenon occurs when vibrations in one object cause vibrations in a second object?

(A) reflection

(B) resonance

(C) intensity

(D) tuning

10.29 A: (B) resonance.

Interference

When more than one wave travels through the same location in the same medium at the same time, the total displacement of the medium is governed by the principle of **superposition**. The principle of superposition simply states that the total displacement is the sum of all the individual displacements of the waves. The combined effect of the interaction of the multiple waves is known as **wave interference**.

10.30 Q: The diagram below shows two pulses approaching each other in a uniform medium. Diagram the superposition of the two pulses.

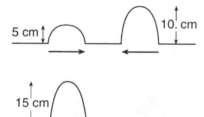

10.30 A: 15 cm

When two or more pulses with displacements in the same direction interact, the effect is known as **constructive interference**. The resulting displacement is greater than the original individual pulses. Once the pulses have passed by each other, they continue along their original path in their original shape, as if they had never met.

When two or more pulses with displacements in opposite directions interact, the effect is known as **destructive interference**. The resulting displacements negate each other. Once the pulses have passed by each other, they continue along their original path in their original shape, as if they had never met. An animation of two pulses interfering constructively and destructively is available at http://bit.ly/hyJ3lZ.

10.31 Q: The diagram below represents two pulses approaching each other from opposite directions in the same medium.

Which diagram best represents the medium after the pulses have passed through each other?

(1)

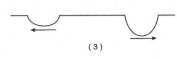

(2)

(3)

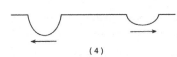

(4)

10.31 A: (2) the pulses continue as if they had never met.

10.32 Q: The diagram below represents shallow water waves of constant wavelength passing through two small openings, A and B, in a barrier.

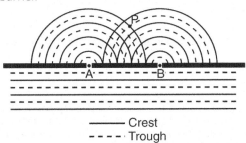

——— Crest
- - - - Trough

Which statement best describes the interference at point P?

(A) It is constructive, and causes a longer wavelength.

(B) It is constructive, and causes an increase in amplitude.

(C) It is destructive, and causes a shorter wavelength.

(D) It is destructive, and causes a decrease in amplitude.

10.32 A: (D) when a crest and a trough meet, destructive interference causes a decrease in amplitude.

10.33 Q: The diagram below shows two pulses of equal amplitude, A, approaching point P along a uniform string.

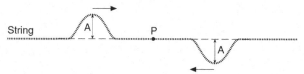

When the two pulses meet at P, the vertical displacement of the string at P will be

(A) A

(B) 2A

(C) 0

(D) A/2

10.33 A: (C) the pulses will experience destructive interference.

10.34 Q: The diagram below represents two pulses approaching each other.

Which diagram best represents the resultant pulse at the instant the pulses are passing through each other?

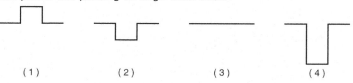

(1)   (2)   (3)   (4)

10.34 A: (2) shows the superposition (addition) of the two pulses.

10.35 Q: Two waves having the same amplitude and frequency are traveling in the same medium. Maximum destructive interference will occur when the phase difference between the waves is

(A) 0°

(B) 90°

(C) 180°

(D) 270°

10.35 A: (C) Maximum destructive interference occurs at a phase difference of 180°.

Beats

When two sound waves of similar (but not exactly the same) frequency approach your ear, they interfere with each other to create an alternating loud and quiet sound. This is a result of alternating constructive and destructive interference.

Consider a wave of frequency 4Hz. A plot of the wave's amplitude as a function of time is shown below.

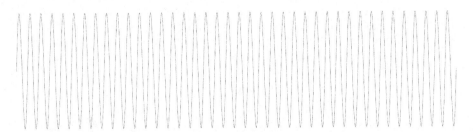

Now, consider a wave of slightly higher frequency, say 5 Hz, as depicted below.

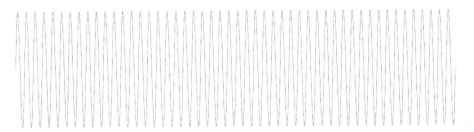

When these two waves interfere, the superposition principle dictates that the amplitude of the waves will add up at each point along the wave. The graph below shows the resulting wave, obtained by taking the sum of the 4Hz and 5Hz waves.

Notice the alternating regions of high and low amplitude corresponding to regions of constructive and destructive interference. This leads to a wave that has an overall oscillation in its amplitude, where the frequency of amplitude oscillations is equal to the difference in frequency of the initial two waves, as highlighted by the "enveloping" wave below.

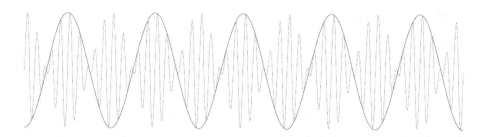

In this manner, the difference in the frequency of two waves can be determined by measuring the frequency of the resulting "beating" wave. The number of amplitude "beats" per second, also known as the **beat frequency**, is equal to the magnitude of the difference in frequency of the two initial waves!

$$f_{beat} = \left| f_2 - f_1 \right|$$

The "beating" phenomena is often used in the tuning of instruments such as a guitar. The guitarist will play what should be the same note (and therefore the same frequency sound) on two separate strings. If beats are heard, the tension in one of the strings is adjusted to minimize the beat frequency until the beating completely disappears, indicating the strings are now creating sound at the same frequency, and are therefore in tune. This is especially helpful given that human ears can typically perceive beat frequencies of 1 to 4 Hz, but are considerably less adept at differentiating tone, struggling to detect differences of 9 Hz at a sound wave frequency of 3 kHz.

10.36 Q: While tuning her guitar, Sandra plays an A note on the fifth string at exactly 110 Hz. Sandra then frets the sixth string to produce the same tone. However, when she frets and plays the sixth string, she hears a beat pattern where she hears the loudest amplitude two times each second. Assuming the 5th string is perfectly in tune, which of the following could be the frequency of the out-of-tune sixth string? Choose all that apply.

(A) 108 Hz

(B) 109.5 Hz

(C) 110.5 Hz

(D) 112 Hz

10.36 A: (A) and (D) are possible frequencies for the out-of-tune string. If the beat frequency is 2 Hz, the out-of-tune string must be two hertz higher or two hertz lower than the in-tune string.

Standing Waves

When waves of the same frequency and amplitude traveling in opposite directions meet, a standing wave is produced. A **standing wave** is a wave in which certain points (**nodes**) appear to be standing still and other points (**antinodes**) vibrate with maximum amplitude above and below the axis.

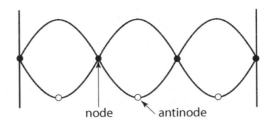

Looking at the standing wave produced above, you can see a total of four nodes in the wave, and three antinodes. For any standing wave pattern, you will always have one more node than antinode.

Standing waves can be observed in a variety of patterns and configurations, and are responsible for the functioning of most musical instruments. Guitar strings, for example, demonstrate a standing wave pattern. By fretting the strings, you adjust the wavelength of the string, and therefore the frequency of the standing wave pattern, creating a different pitch. Similar functionality is seen in instruments ranging from pianos and drums to flutes, harps, trombones, xylophones, and even pipe organs!

10.37 Q: While playing, two children create a standing wave in a rope, as shown in the diagram below.

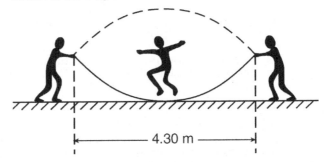

A third child participates by jumping the rope. What is the wavelength of this standing wave?

(A) 2.15 m

(B) 4.30 m

(C) 6.45 m

(D) 8.60 m

10.37 A: (D) the standing wave shown is half a wavelength, therefore the total wavelength must be 8.6m.

10.38 Q: Wave X travels eastward with frequency f and amplitude A. Wave Y, traveling in the same medium, interacts with wave X and produces a standing wave. Which statement about wave Y is correct?

(A) Wave Y must have a frequency of f, an amplitude of A, and be traveling eastward.

(B) Wave Y must have a frequency of 2f, an amplitude of 3A, and be traveling eastward.

(C) Wave Y must have a frequency of 3f, an amplitude of 2A, and be traveling westward.

(D) Wave Y must have a frequency of f, an amplitude of A, and be traveling westward.

10.38 A: (D) Standing waves are created when waves with the same frequency and amplitude traveling in opposite directions meet.

10.39 Q: The diagram below represents a wave moving toward the right side of this page.

Which wave shown below could produce a standing wave with the original wave?

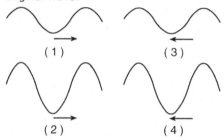

(1)   (3)

(2)   (4)

10.39 A: (3) must have same frequency, amplitude, and be traveling in the opposite direction in the same medium.

10.40 Q: Rank the following four standing waves in the same medium from highest to lowest in terms of A) frequency, B) period, C) wavelength, and D) amplitude.

A   B

C   D

10.40 A: A) Frequency: D, C, B, A

B) Period: A, B, C, D

C) Wavelength: A, B, C, D

D) Amplitude: C, B, D, A

10.41 Q: The diagram below shows a standing wave.

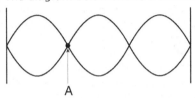

Point A on the standing wave is

(A) a node resulting from constructive interference

(B) a node resulting from destructive interference

(C) an antinode resulting from constructive interference

(D) an antinode resulting from destructive interference

10.41 A: (B) a node resulting from destructive interference.

10.42 Q: One end of a rope is attached to a variable speed drill and the other end is attached to a 5-kilogram mass. The rope is draped over a hook on a wall opposite the drill. When the drill rotates at a frequency of 20 Hz, standing waves of the same frequency are set up in the rope. The diagram below shows such a wave pattern.

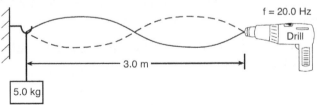

A) Determine the wavelength of the waves producing the standing wave pattern.

B) Calculate the speed of the wave in the rope.

10.42 A: A) Wavelength is 3.0 meters from diagram.

B) $v = f\lambda = (20\,Hz)(3m) = 60\,{}^{m}\!/_{s}$

Standing Waves in Instruments

One method of creating standing waves involves holding a string in place at both ends and introducing a disturbance. Because the ends are held in place, the ends must serve as nodes. This allows only certain wavelengths within the string, making this a popular method of creating musical instruments. In fact, many stringed instruments utilize this principle, including the piano, bass, cello, violin, and guitar!

The **fundamental frequency** is created by holding a string at both ends, with half a wavelength set up between the nodes. This frequency is also known as the **first harmonic**. If, instead, a full wavelength is introduced between the nodes, the first overtone, also known as the second harmonic, is produced. If one and a half wavelengths are introduced between the nodes, the second overtone, or 3rd harmonic, is produced. It is the combination of fundamental frequencies and overtones which give various instruments their characteristic sounds, which is why a 440-Hz "A" played on a piano sounds different than a 440-Hz "A" played on a violin.

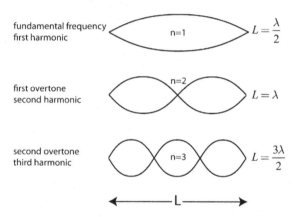

From these principles, you can develop a basic set of relationships between the harmonic, the length of the string, the wavelength, and the resulting frequency.

First, note that the length of the string is always half the harmonic number multiplied by the wavelength.

$$L = \frac{n\lambda}{2} \rightarrow \lambda = \frac{2L}{n}$$

With just a bit of manipulation and the addition of the wave equation, you can find the frequency of each harmonic as a function of the wave speed and the length of the string!

$$\lambda = \frac{2L}{n} \xrightarrow[\lambda = \frac{v}{f}]{v = f\lambda} \frac{2L}{n} = \frac{v}{f} \rightarrow f = \frac{nv}{2L}$$

10.43 Q: A guitar designer is designing an instrument in which the speed of the waves on a string should be 450 m/s. How long must the string be to produce a first fundamental frequency F# note at 370 Hz?

10.43 A: $f = \dfrac{nv}{2L} \rightarrow L = \dfrac{nv}{2f} = \dfrac{(1)450\,m/_s}{2(370Hz)} = 0.61m$

10.44 Q: Which frequency or frequencies could not be produced by a string of length 1m given a wave speed of 450 m/s across the string? Choose all that apply.

(A) 225 Hz

(B) 450 Hz

(C) 750 Hz

(D) 900 Hz

10.44 A: (C) 750 Hz. 225 Hz is the fundamental frequency, or first harmonic, 450 Hz is the second harmonic, and 900 Hz is the fourth harmonic. The third harmonic would be 675 Hz, but is not one of the available choices.

Other types of instruments create sound through the formation of standing waves in open and closed tubes. The wavelengths of the sound waves, and therefore the frequencies produced, are determined by the size of the tube confining the wave. Keeping in mind that air is what is moving in these tubes, the air at the closed end of a tube moves with minimum displacement, forming a node. Air at the open end of a tube moves with maximum displacement, forming an antinode.

Instruments using tubes which are open at both ends, such as trumpets, pipe organs, flutes, clarinets, oboes, and others, must have an antinode at each end of the tube. You can perform an analysis similar to that used on stringed instruments as shown below.

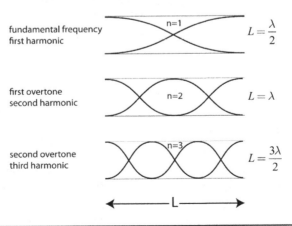

fundamental frequency
first harmonic
n=1
$L = \dfrac{\lambda}{2}$

first overtone
second harmonic
n=2
$L = \lambda$

second overtone
third harmonic
n=3
$L = \dfrac{3\lambda}{2}$

L

Similar to strings, the length of the tube is always half the harmonic number multiplied by the wavelength, and the frequencies produced are determined in the same manner.

$$\lambda = \frac{2L}{n} \qquad f = \frac{nv}{2L}$$

Certain instruments such as the clarinet and some pipe organs use a tube which is open at only one end. Closed-end tubes must have a node at the closed end and an antinode at the open end. This limits the instrument to forming only the fundamental frequency and odd harmonics, as shown below.

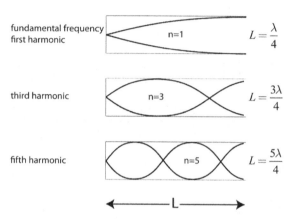

fundamental frequency
first harmonic
$n=1$
$L = \frac{\lambda}{4}$

third harmonic
$n=3$
$L = \frac{3\lambda}{4}$

fifth harmonic
$n=5$
$L = \frac{5\lambda}{4}$

$\longleftarrow$ L $\longrightarrow$

For these situations, the equations for the wavelengths produced and the frequencies change slightly. Noting that with only the fundamental frequency and odd harmonics allowed, n can only equal odd integer values. You can therefore write the wavelength and frequency equations for a tube closed at one end as:

$$\lambda = \frac{4L}{n} \qquad f = \frac{nv}{4L}$$

10.45 Q: The diagram below shows air displacement of four standing waves inside a set of organ pipes. Assume the velocity of sound in air is 343 m/s.

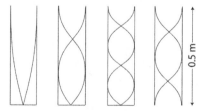

0.5 m

A) What is the highest frequency for the waves shown?
B) What is the lowest frequency for the waves shown?
C) What is the longest wavelength shown in the pipes?
D) What is the shortest wavelength shown in the pipes?

10.45 A: A) 858 Hz (frequency of 3rd pipe, which is a 5th harmonic)

B) 172 Hz (frequency of 1st pipe, which is the fundamental)

C) 2m (length of wave in 1st pipe)

D) 0.4 m (length of wave in 3rd pipe)

10.46 Q: A musician is designing a custom instrument which utilizes a tube open at both ends. Given the speed of sound in air is 343 m/s, how long should the musician make the tube to create an A (440 Hz) as the instrument's fundamental frequency?

10.46 A: $f = \dfrac{nv}{2L} \rightarrow L = \dfrac{nv}{2f} = \dfrac{1(343\,m/s)}{2(440Hz)} = 0.39m$

Doppler Effect

The shift in a wave's observed frequency due to relative motion between the source of the wave and the observer is known as the **Doppler Effect**. In essence, when the source and/or observer are moving toward each other, the observer perceives a shift to a higher frequency, and when the source and/or observer are moving away from each other, the observer perceives a lower frequency.

This can be observed when a vehicle travels past you. As you hear the vehicle approach, you can observe a higher frequency noise, and as the vehicle passes by you and then moves away, you observe a lower frequency noise. This effect is the principle behind radar guns to measure an object's speed as well as meteorology radar which provides data on wind speeds.

The Doppler Effect results from waves having a fixed speed in a given medium. As waves are emitted, a moving source or observer encounters the wave fronts at a different frequency than the waves are emitted, resulting in a perceived shift in frequency. The video and animation at http://bit.ly/epLkPj may help you visualize this effect.

10.47 Q: A car's horn produces a sound wave of constant frequency. As the car speeds up going away from a stationary spectator, the sound wave detected by the spectator

(A) decreases in amplitude and decreases in frequency

(B) decreases in amplitude and increases in frequency

(C) increases in amplitude and decreases in frequency

(D) increases in amplitude and increases in frequency

10.47 A: (A) decreases in amplitude because the distance between source and observe is increasing, and decreases in frequency because the source is moving away from the observer.

10.48 Q: A car's horn is producing a sound wave having a constant frequency of 350 hertz. If the car moves toward a stationary observer at constant speed, the frequency of the car's horn detected by this observer may be

(A) 320 Hz

(B) 330 Hz

(C) 350 Hz

(D) 380 Hz

10.48 A: (D) If source is moving toward the stationary observer, the observed frequency must be higher than source frequency.

10.49 Q: A radar gun can determine the speed of a moving automobile by measuring the difference in frequency between emitted and reflected radar waves. This process illustrates

(A) resonance

(B) the Doppler effect

(C) diffraction

(D) refraction

10.49 A: (B) the Doppler effect.

Chapter 10: Mechanical Waves

10.50 Q: The vertical lines in the diagram represent compressions in a sound wave of constant frequency propagating to the right from a speaker toward an observer at point A.

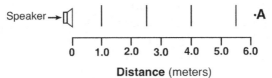

A) Determine the wavelength of this sound wave.

B) The speaker is then moved at constant speed toward the observer at A. Compare the wavelength of the sound wave received by the observer while the speaker is moving to the wavelength observed when the speaker was at rest.

10.50 A: A) Wavelength is compression to compression, or 1.5m.

B) Observed frequency is higher while speaker is moving toward the observer due to the Doppler Effect, so the observed wavelength must be shorter.

10.51 Q: A student sees a train that is moving away from her and sounding its whistle at a constant frequency. Compared to the sound produced by the whistle, the sound observed by the student is

(A) greater in amplitude

(B) a transverse wave rather than a longitudinal wave

(C) higher in pitch

(D) lower in pitch

10.51 A: (D) lower in pitch since the source is moving away from the observer.

An exciting application of the Doppler Effect involves the analysis of radiation from distant stars and galaxies in the universe. Based on the basic elements that compose stars, scientists know what frequencies of radiation to look for. However, when analyzing these objects, they observe frequencies shifted toward the red end of the electromagnetic spectrum (lower frequencies), known as the **Red Shift**. This indicates that these celestial objects must be moving away from the Earth. The more distant the object, the greater the red shift. Putting this together, you can conclude that more distant celestial objects are moving away from Earth faster than nearer objects, and therefore, the universe must be expanding!

10.52 Q: When observed from Earth, the wavelengths of light emitted by a star are shifted toward the red end of the electromagnetic spectrum. This Red Shift occurs because the star is

(A) at rest relative to Earth

(B) moving away from Earth

(C) moving toward Earth at decreasing speed

(D) moving toward Earth at increasing speed

10.52 A: (B) moving away from Earth.

Test Your Understanding

1. List at least five examples illustrating how waves transfer energy.

2. Explain why singers attempting to shatter glass tend to use high quality crystal glasses instead of mason jars.

3. What determines the speed of a wave?

4. Research noise-canceling headphones. Explain how they function using the concept of interference.

5. Derive the equations for the wavelengths and frequencies of a tube which is closed at one end.

6. In your own words, explain why open tubes produce all harmonics, while closed tubes produce only odd harmonics.

7. Does a radar gun's accuracy depend on the angle between the moving object and the radar gun? Explain.

8. Why do you sound so great (to yourself) while singing in the shower?

Chapter 11: Electrostatics

"Electricity can be dangerous. My nephew tried to stick a penny into a plug. Whoever said a penny doesn't go far didn't see him shoot across that floor.

I told him he was grounded."

— Tim Allen

Objectives

1. Describe the smallest isolated electric charge as an elementary charge.
2. Explain the charge on a proton or neutron as a result of the particle's quark composition.
3. Calculate the charge on an object.
4. Describe the differences between conductors and insulators.
5. Solve problems using the law of conservation of charge.
6. Use Coulomb's Law to solve problems related to electrical force.
7. Recognize that objects that are charged exert forces, both attractive and repulsive.
8. Compare and contrast Newton's Law of Universal Gravitation with Coulomb's Law.
9. Define, measure, and calculate the strength of an electric field.
10. Solve problems related to charge, electric field, and forces.
11. Define and calculate electric potential energy.
12. Define and calculate potential difference.
13. Solve basic problems involving charged parallel plates.

Electricity and magnetism play a profound role in almost all aspects of life. From the moment you wake up, to the moment you go to sleep (and even while you're sleeping), applications of electricity and magnetism provide tools, light, warmth, transportation, communication, and even entertainment. Despite its widespread use, however, there is much about these phenomena that is not well understood.

Electric Charges

Matter is made up of atoms. Once thought to be the smallest building blocks of matter, scientists now know that atoms can be broken up into even smaller pieces, known as protons, electrons, and neutrons. Each atom consists of a dense core of positively charged protons and uncharged (neutral) neutrons. This core is known as the nucleus. It is surrounded by a "cloud" of much smaller, negatively charged electrons. These electrons orbit the nucleus in distinct energy levels. To move to a higher energy level, an electron must absorb energy. When an electron falls to a lower energy level, it gives off energy.

Most atoms are neutral -- that is, they have an equal number of positive and negative charges, giving a net charge of 0. In certain situations, however, an atom may gain or lose electrons. In these situations, the atom as a whole is no longer neutral and is called an **ion**. If an atom loses one or more electrons, it has a net positive charge and is known as a positive ion. If, instead, an atom gains one or more electrons, it has a net negative charge and is therefore called a negative ion. Like charges repel each other, while opposite charges attract each other. In physics, the charge on an object is represented with the symbol q.

Charge is a fundamental measurement in physics, much as length, time, and mass are fundamental measurements. The fundamental unit of charge is the **coulomb** [C], which is a very large amount of charge. Compare that to the magnitude of charge on a single proton or electron, known as an elementary charge (e), which is equal to 1.6×10^{-19} coulomb. It would take 6.25×10^{18} elementary charges to make up a single coulomb of charge!

11.01 Q: An object possessing an excess of 6.0×10^6 electrons has what net charge?

11.01 A: 6×10^6 electrons $\bullet \dfrac{-1.6 \times 10^{-19} C}{1 \text{ electron}} = -9.6 \times 10^{-13} C$

11.02 Q: An alpha particle consists of two protons and two neutrons. What is the charge of an alpha particle?

(A) 1.25×10^{19} C

(B) 2.00 C

(C) 6.40×10^{-19} C

(D) 3.20×10^{-19} C

11.02 A: (D) The net charge on the alpha particle is +2 elementary charges.

$$2e \bullet \frac{1.6 \times 10^{-19} C}{1e} = 3.2 \times 10^{-19} C$$

11.03 Q: If an object has a net negative charge of 4 coulombs, the object possesses

(A) 6.3×10^{18} more electrons than protons

(B) 2.5×10^{19} more electrons than protons

(C) 6.3×10^{18} more protons than electrons

(D) 2.5×10^{19} more protons than electrons

11.03 A: (B) $-4C \bullet \dfrac{1e}{1.6 \times 10^{-19} C} = -2.5 \times 10^{19} e$

11.04 Q: Which quantity of excess electric charge could be found on an object?

(A) 6.25×10^{-19} C

(B) 4.80×10^{-19} C

(C) 6.25 elementary charges

(D) 1.60 elementary charges

11.04 A: (B) all other choices require fractions of an elementary charge, while choice (B) is an integer multiple (3e) of elementary charges.

11.05 Q: What is the net electrical charge on a magnesium ion that is formed when a neutral magnesium atom loses two electrons?

(A) -3.2×10^{-19} C

(B) -1.6×10^{-19} C

(C) $+1.6 \times 10^{-19}$ C

(D) $+3.2 \times 10^{-19}$ C

11.05 A: (D) the net charge must be +2e, or $2(1.6 \times 10^{-19}$ C$)=3.2 \times 10^{-19}$ C.

The Standard Model

Where does electric charge come from? To answer that question, you need to dive into the **Standard Model of Particle Physics**. As you've learned previously, the atom is the smallest part of an element (such as oxygen) that has the characteristics of the element. Atoms are made up of very small negatively charged electrons surrounding the much larger nucleus. The nucleus is composed of positively charged protons and neutral neutrons. The positively charged protons exert a repelling electrical force upon each other, but the strong nuclear force holds the protons and neutrons together in the nucleus.

This completely summarized scientists' understanding of atomic structure until the 1930s, when scientists began to discover evidence that there was more to the picture and that protons and nucleons were made up of even smaller particles. This launched the particle physics movement, which, to this day, continues to challenge the understanding of the entire universe by exploring the structure of the atom.

In addition to standard matter, researchers have discovered the existence of antimatter. **Antimatter** is matter made up of particles with the same mass as regular matter particles, but opposite charges and other characteristics. An **antiproton** is a particle with the same mass as a proton, but a negative (opposite) charge. A **positron** has the same mass as an electron, but a positive charge. An **antineutron** has the same mass as a neutron, but has other characteristics opposite that of the neutron.

When a matter particle and its corresponding antimatter particle meet, the particles may combine to **annihilate** each other, resulting in the complete conversion of both particles into energy.

This book has dealt with many types of forces, ranging from contact forces such as tensions and normal forces to field forces such as the electrical force and gravitational force. When observed from their most basic aspects, however, all observed forces in the universe can be consolidated into four fundamental forces. They are, from strongest to weakest:

1. **Strong Nuclear Force**: holds protons and neutrons together in the nucleus
2. **Electromagnetic Force**: electrical and magnetic attraction and repulsion
3. **Weak Nuclear Force**: responsible for radioactive beta decay
4. **Gravitational Force**: attractive force between objects with mass

Understanding these forces remains a topic of scientific research, with current work exploring the possibility that forces are actually conveyed by an exchange of force-carrying particles such as photons, bosons, gluons, and gravitons.

11.06 Q: The particles in a nucleus are held together primarily by the
(A) strong force
(B) gravitational force
(C) electrostatic force
(D) magnetic force

11.06 A: (A) the strong nuclear force holds protons and neutrons together in the nucleus.

11.07 Q: Which statement is true of the strong nuclear force?
(A) It acts over very great distances.
(B) It holds protons and neutrons together.
(C) It is much weaker than gravitational forces.
(D) It repels neutral charges.

11.07 A: (B) The strong nuclear force holds protons and neutrons together.

11.08 Q: The strong force is the force of
(A) repulsion between protons
(B) attraction between protons and electrons
(C) repulsion between nucleons
(D) attraction between nucleons

11.08 A: (D) attraction between nucleons (nucleons are particles in the nucleus such as protons and neutrons).

The current model of sub-atomic structure used to understand matter is known as the Standard Model. Development of this model began in the late 1960s, and has continued through today with contributions from many scientists across the world. The Standard Model explains the interactions of the strong (nuclear), electromagnetic, and weak forces, but has yet to account for the gravitational force.

Although the Standard Model itself is a very complicated theory, the basic structure of the model is fairly straightforward. According to the model, all matter is divided into two categories, known as **hadrons** and the much smaller **leptons**. All of the fundamental forces act on hadrons, which include particles such as protons and neutrons. In contrast, the strong nuclear force doesn't act on leptons, so only three fundamental forces act on leptons, such as electrons, positrons, muons, tau particles and neutrinos.

Hadrons are further divided into **baryons** and **mesons**. Baryons such as protons and neutrons are composed of three smaller particles known as **quarks**. Charges of baryons are always whole numbers. Mesons are composed of a quark and an anti-quark (for example, an up quark and an anti-down quark).

Classification of Matter

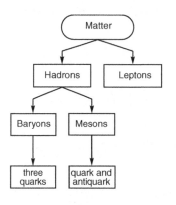

Particles of the Standard Model

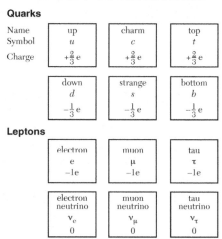

Scientists have identified six types of quarks. For each of the six types of quarks, there also exists a corresponding anti-quark with an opposite charge. The quarks have rather interesting names: up quark, down quark, charm quark, strange quark, top quark, and bottom quark. Charges on each quark are either one third of an elementary charge, or two thirds of an elementary charge, positive or negative, and the quarks are symbolized by the first letter of their name. For the associated anti-quark, the symbol is the first letter of the anti-quark's name, with a line over the name. For example, the symbol for the up quark is u. The symbol for the anti-up quark is ū.

Similarly, scientists have identified six types of leptons: the electron, the muon, the tau particle, and the electron neutrino, muon neutrino, and tau neutrino. Again, for each of these leptons there also exists an associated anti-lepton. The most familiar lepton, the electron, has a charge of -1e. Its anti-particle, the positron, has a charge of +1e.

Since a proton is made up of three quarks, and has a positive charge, the sum of the charges on its constituent quarks must be equal to one elementary charge. A proton is actually comprised of two up quarks and one down quark. You can verify this by adding up the charges of the proton's constituent quarks (uud).

$$\left(+\frac{2}{3}e\right)+\left(+\frac{2}{3}e\right)+\left(-\frac{1}{3}e\right)=+1e$$

Chapter 11: Electrostatics

11.09 Q: A neutron is composed of up and down quarks. How many of each type of quark are needed to make a neutron?

11.09 A: The charge on the neutron must sum to zero, and the neutron is a baryon, so it is made up of three quarks. To achieve a total charge of zero, the neutron must be made up of one up quark (+2/3e) and two down quarks (-1/3e).

If the charge on a quark (such as the up quark) is (+2/3)e, the charge of the anti-quark (ū) is (-2/3)e. The anti-quark is the same type of particle, with the same mass, but with the opposite charge.

11.10 Q: What is the charge of the down anti-quark?

11.10 A: The down quark's charge is -1/3e, so the anti-down quark's charge must be +1/3e.

11.11 Q: Compared to the mass and charge of a proton, an antiproton has
(A) the same mass and the same charge
(B) greater mass and the same charge
(C) the same mass and the opposite charge
(D) greater mass and the opposite charge

11.11 A: (C) the same mass and the opposite charge.

Conductors and Insulators

Certain materials allow electric charges to move freely. These are called **conductors**. Examples of good conductors include metals such as gold, copper, silver, and aluminum. In contrast, materials in which electric charges cannot move freely are known as **insulators**. Good insulators include materials such as glass, plastic, and rubber. Metals are better conductors of electricity compared to insulators because metals contain more free electrons.

Conductors and insulators are characterized by their resistivity, or ability to resist movement of charge. Materials with high resistivities are good insulators. Materials with low resistivities are good conductors.

Semiconductors are materials which, in pure form, are good insulators. However, by adding small amounts of impurities known as dopants, their resistivities can be lowered significantly until they become good conductors.

Charging by Conduction*

Materials can be charged by contact, or **conduction**. If you take a balloon and rub it against your hair, some of the electrons from the atoms in your hair are transferred to the balloon. The balloon now has extra electrons, and therefore has a net negative charge. Your hair has a deficiency of electrons, so therefore it now has a net positive charge.

Much like momentum and energy, charge is also conserved in a closed system. Continuing the hair and balloon example, the magnitude of the net positive charge on your hair is equal to the magnitude of the net negative charge on the balloon. The total charge of the hair/balloon system remains zero (neutral). For every extra electron (negative charge) on the balloon, there is a corresponding missing electron (positive charge) in your hair. This known as the law of conservation of charge.

Conductors can also be charged by contact. If a charged conductor is brought into conduct with an identical neutral conductor, the net charge will be shared across the two conductors.

11.12 Q: If a conductor carrying a net charge of 8 elementary charges is brought into contact with an identical conductor with no net charge, what will be the charge on each conductor after they are separated?

11.12 A: Each conductor will have a charge of 4 elementary charges.

11.13 Q: What is the net charge (in coulombs) on each conductor after they are separated?

11.13 A: $q=4e=4(1.6\times10^{-19}\ C)=6.4\times10^{-19}\ C$

11.14 Q: Metal sphere A has a charge of −2 units and an identical metal sphere, B, has a charge of −4 units. If the spheres are brought into contact with each other and then separated, the charge on sphere B will be

(A) 0 units

(B) -2 units

(C) -3 units

(D) +4 units

11.14 A: (C) -3 units.

11.15 Q: Compared to insulators, metals are better conductors of electricity because metals contain more free

(A) protons

(B) electrons

(C) positive ions

(D) negative ions

11.15 A: (B) electrons.

A simple tool used to detect small electric charges known as an **electroscope** functions on the basis of conduction. The electroscope consists of a conducting rod attached to two thin conducting leaves at one end and isolated from surrounding charges by an insulating stopper placed in a flask. If a charged object is placed in contact with the conducting rod, part of the charge is transferred to the rod. Because the rod and leaves form a conducting path and like charges repel each other, the charges are distributed equally along the entire rod and leaf apparatus. The leaves, having like charges, repel each other, with larger charges providing greater leaf separation!

11.16 Q: Separation of the leaves of an electroscope when an object is touched to the electroscope's conducting rod could indicate (choose all that apply):

(A) the object is neutral

(B) the object is negatively charged

(C) the object is positively charged

(D) the object is an insulator

11.16 A: (B & C) Separation of the leaves of an electroscope indicate the object touching the electroscope is charge. Whether the charge is positive or negative cannot be determined with just this test.

Charging by Induction*

Conductors can also be charged without coming into contact with another charged object in a process known as charging by **induction**. This is accomplished by placing the conductor near a charged object and grounding the conductor. To understand charging by induction, you must first realize that when an object is connected to the earth by a conducting path, known as grounding, the earth acts like an infinite source for providing or accepting excess electrons.

To charge a conductor by induction, you first bring it close to another charged object. When the conductor is close to the charged object, any free electrons on the conductor will move toward the charged object if the object is positively charged (since opposite charges attract) or away from the charged object if the object is negatively charged (since like charges repel).

If the conductor is then "grounded" by means of a conducting path to the Earth, the excess charge is compensated for by means of electron transfer to or from earth. Then the ground connection is severed. When the originally charged object is moved far away from the conductor, the charges in the conductor redistribute, leaving a net charge on the conductor as shown.

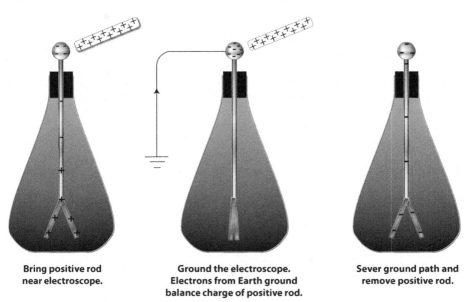

Bring positive rod near electroscope.

Ground the electroscope. Electrons from Earth ground balance charge of positive rod.

Sever ground path and remove positive rod.

You can also induce a charge in a charged region in a neutral object by bringing a strong positive or negative charge close to that object. In such cases, the electrons in the neutral object tend to move toward a strong positive charge, or away from a large negative charge. Though the object itself remains neutral, portions of the object are more positive or negative than other parts. In this way, you can attract a neutral object by bringing a charged object close to it, positive or negative. Put another way, a positively charged object can be attracted to both a negatively charged object and a neutral object, and a negatively charged object can be attracted to both a positively charged object and a neutral object.

For this reason, the only way to tell if an object is charged is by repulsion. A positively charge object can only be repelled by another positive charge and a negatively charged object can only be repelled by another negative charge.

11.17 Q: A positively charged glass rod attracts object X. The net charge of object X

(A) may be zero or negative

(B) may be zero or positive

(C) must be negative

(D) must be positive

11.17 A: (A) a positively charged rod can attract a neutral object or a negatively charged object.

11.18 Q: The diagram below shows three neutral metal spheres, x, y, and z, in contact and on insulating stands.

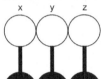

Which diagram best represents the charge distribution on the spheres when a positively charged rod is brought near sphere x, but does not touch it?

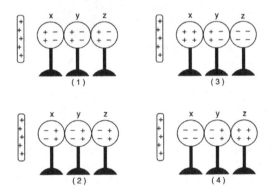

11.18 A: (4) is the correct answer.

Coulomb's Law

Like charges repel and opposite charges attract. In order for charges to repel or attract, they apply a force upon each either, known as the **electrostatic force**. Similar to the manner in which the force of attraction between two masses is determined by the amount of mass and the distance between the masses, as described by Newton's Law of Universal Gravitation, the force of attraction or repulsion is determined by the amount of charge and the distance between the charges.

The magnitude of the electrostatic force is described by **Coulomb's Law**, which states that the magnitude of the electrostatic force (F_e) between two objects is equal to a constant, k, multiplied by each of the two charges, q_1 and q_2, and divided by the square of the distance between the charges (r^2). The constant k is known as the **electrostatic constant** and is given as $k=8.99\times10^9$ N·m²/C².

$$F_e = \frac{kq_1q_2}{r^2}$$

Notice how similar this formula is to the formula for the gravitational force! Both Newton's Law of Universal Gravitation and Coulomb's Law follow the inverse-square relationship, a pattern that repeats many times over in physics. The further you get from the charges, the weaker the electrostatic force. If you were to double the distance from a charge, you would quarter the electrostatic force on a charge.

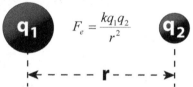

Formally, a positive value for the electrostatic force indicates that the force is a repelling force, while a negative value for the electrostatic force indicates that the force is an attractive force. Because force is a vector, you must assign a direction to it. To determine the direction of the force vector, once you have calculated its magnitude, use common sense to tell you the direction on each charged object. If the objects have opposite charges, they attract each other, and if they have like charges, they repel each other.

11.19 Q: Three protons are separated from a single electron by a distance of 1×10^{-6} m. Find the electrostatic force between them. Is this force attractive or repulsive?

11.19 A: $q_1 = 3$ protons $= 3(1.6\times10^{-19}C) = 4.8\times10^{-19}C$

$q_2 = 1$ electron $= 1(-1.6\times10^{-19}C) = -1.6\times10^{-19}C$

$F_e = \frac{kq_1q_2}{r^2} = \frac{(8.99\times10^9\ \frac{N\bullet m^2}{C^2})(4.8\times10^{-19}C)(-1.6\times10^{-19}C)}{(1\times10^{-6}m)^2}$

$F_e = -6.9\times10^{-16}N$ attractive

11.20 Q: A distance of 1.0 meter separates the centers of two small charged spheres. The spheres exert gravitational force F_g and electrostatic force F_e on each other. If the distance between the spheres' centers is increased to 3.0 meters, the gravitational force and electrostatic force, respectively, may be represented as

(A) $F_g/9$ and $F_e/9$

(B) $F_g/3$ and $F_e/3$

(C) $3F_g$ and $3F_e$

(D) $9F_g$ and $9F_e$

11.20 A: (A) due to the inverse square law relationships.

11.21 Q: A beam of electrons is directed into the electric field between two oppositely charged parallel plates, as shown in the diagram below.

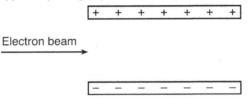

Electron beam

The electrostatic force exerted on the electrons by the electric field is directed

(A) into the page

(B) out of the page

(C) toward the bottom of the page

(D) toward the top of the page

11.21 A: (D) toward the top of the page because the electron beam is negative, and will be attracted by the positively charged upper plate and repelled by the negatively charged lower plate.

11.22 Q: The centers of two small charged particles are separated by a distance of 1.2×10^{-4} meter. The charges on the particles are $+8.0\times10^{-19}$ coulomb and $+4.8\times10^{-19}$ coulomb, respectively.

(A) Calculate the magnitude of the electrostatic force between these two particles.

(B) Sketch a graph showing the relationship between the magnitude of the electrostatic force between the two charged particles and the distance between the centers of the particles.

11.22 A: (A) $F_e = \dfrac{kq_1q_2}{r^2} = \dfrac{(8.99\times10^9\,\frac{N\bullet m^2}{C^2})(8.0\times10^{-19}\,C)(4.8\times10^{-19}\,C)}{(1.2\times10^{-4}\,m)^2}$

$F_e = 2.4\times10^{-19}\,N$

(B)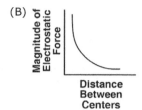

11.23 Q: The diagram below shows a beam of electrons fired through the region between two oppositely charged parallel plates in a cathode ray tube.

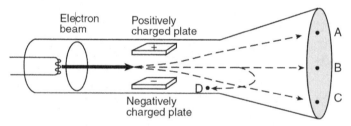

After passing between the charged plates, the electrons will most likely travel path

(A) A
(B) B
(C) C
(D) D

11.23A: (A) A

Electric Fields*

Similar to gravity, the electrostatic force is a non-contact force, or field force. Charged objects do not have to be in contact with each other to exert a force on each other. Somehow, a charged object feels the effect of another charged object through space. The property of space that allows a charged object to feel a force is a concept called the electric field. Although you cannot see an electric field, you can detect its presence by placing a positive test charge at various points in space and measuring the force the test charge feels.

While looking at gravity, the gravitational field strength was the amount of force observed by a mass per unit mass. In similar fashion, the electric field strength is the amount of electrostatic force observed by a charge per unit charge. Therefore, the electric field strength, E, is the electrostatic force observed at a given point in space divided by the test charge itself. Electric field strength is measured in Newtons per Coulomb (N/C).

$$E = \frac{F_e}{q}$$

11.24 Q: Two oppositely charged parallel metal plates, 1.00 centimeter apart, exert a force with a magnitude of 3.60 × 10⁻¹⁵ newtons on an electron placed between the plates. Calculate the magnitude of the electric field strength between the plates.

11.24 A: $E = \dfrac{F_e}{q} = \dfrac{3.6 \times 10^{-15}\,N}{1.6 \times 10^{-19}\,C} = 2.25 \times 10^4\,{}^{N}\!/_{C}$

11.25 Q: Which quantity and unit are correctly paired?
(A) resistivity and Ω/m
(B) potential difference and eV
(C) current and C•s
(D) electric field strength and N/C

11.25 A: (D) electric field strength and N/C.

11.26 Q: What is the magnitude of the electric field intensity at a point where a proton experiences an electrostatic force of magnitude 2.30×10⁻²⁵ newtons?
(A) 3.68×10⁻⁴⁴ N/C
(B) 1.44×10⁻⁶ N/C
(C) 3.68×10⁶ N/C
(D) 1.44×10⁴⁴ N/C

11.26 A: (B) $E = \dfrac{F_e}{q} = \dfrac{2.3 \times 10^{-25}\,N}{1.6 \times 10^{-19}\,C} = 1.44 \times 10^{-6}\,{}^{N}\!/_{C}$

11.27 Q: The diagram below represents an electron within an electric field between two parallel plates that are charged with a potential difference of 40.0 volts.

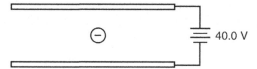

If the magnitude of the electric force on the electron is 2.00×10^{-15} newtons, the magnitude of the electric field strength between the charged plates is

(A) 3.20×10^{-34} N/C

(B) 2.00×10^{-14} N/C

(C) 1.25×10^4 N/C

(D) 2.00×10^{16} N/C

11.27 A: (C) $E = \dfrac{F_e}{q} = \dfrac{2 \times 10^{-15}\,N}{1.6 \times 10^{-19}\,C} = 1.25 \times 10^4 \;{}^N\!/_C$

Since you can't actually see the electric field, you can draw electric field lines to help visualize the force a charge would feel if placed at a specific position in space. These lines show the direction of the electric force a positively charged particle would feel at that point. The more dense the lines are, the stronger the force a charged particle would feel, therefore the stronger the electric field. As the lines get further apart, the strength of the electric force a charged particle would feel is smaller, therefore the electric field is smaller.

By convention, electric field lines are drawn showing the direction of force on a positive charge. Therefore, to draw electric field lines for a system of charges, follow these basic rules:

1. Electric field lines point away from positive charges and toward negative charges.
2. Electric field lines never cross.
3. Electric field lines always intersect conductors at right angles to the surface.
4. Stronger fields have closer lines.
5. Field strength and line density decreases as you move away from the charges.

Let's take a look at a few examples of electric field lines, starting with isolated positive (left) and negative (right) charges. Notice that for each charge, the lines radiate outward or inward spherically. The lines point away from the positive charge, since a positive test charge placed in the field (near the fixed charge) would feel a repelling force. The lines point in toward the negative fixed charge, since a positive test charge would feel an attractive force.

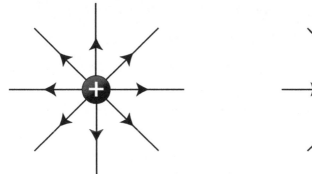

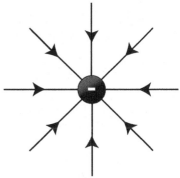

If you have both positive and negative charges in close proximity, you follow the same basic procedure:

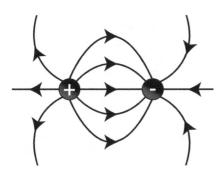

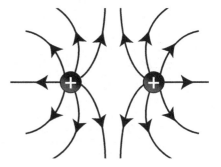

11.28 Q: Two small metallic spheres, A and B, are separated by a distance of 4.0×10^{-1} meter, as shown. The charge on each sphere is $+1.0 \times 10^{-6}$ coulomb. Point P is located near the spheres.

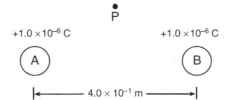

What is the magnitude of the electrostatic force between the two charged spheres?

(A) 2.2×10^{-2} N

(B) 5.6×10^{-2} N

(C) 2.2×10^{4} N

(D) 5.6×10^{4} N

11.28 A: (B) $F_e = \dfrac{kq_1 q_2}{r^2} = \dfrac{(8.99 \times 10^9 \frac{N \bullet m^2}{C^2})(1.0 \times 10^{-6} C)(1.0 \times 10^{-6} C)}{(4 \times 10^{-1} m)^2}$

$F_e = 0.056 N$

11.29 Q: In the diagram below, P is a point near a negatively charged sphere.

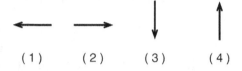

Which vector best represents the direction of the electric field at point P?

11.29 A: Correct answer is (1). Electric field lines point in toward negative charges.

11.30 Q: Sketch at least four electric field lines with arrowheads that represent the electric field around a negatively charged conducting sphere.

11.30 A:

11.31 Q: The centers of two small charged particles are separated by a distance of 1.2×10^{-4} meter. The charges on the particles are $+8.0 \times 10^{-19}$ coulomb and $+4.8 \times 10^{-19}$ coulomb, respectively. Sketch at least four electric field lines in the region between the two positively charged particles.

11.31 A:

Note that these field lines are not symmetrical since the charges have differing magnitudes.

11.32 Q: Which graph best represents the relationship between the magnitude of the electric field strength, E, around a point charge and the distance, r, from the point charge?

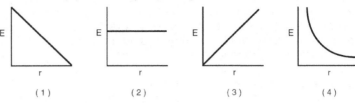

(1)   (2)   (3)   (4)

11.32 A: Correct answer is 4.

Because gravity and electrostatics have so many similarities, let's take a minute to do a quick comparison of electrostatics and gravity.

Electrostatics	Gravity
Force: $F_e = \dfrac{kq_1q_2}{r^2}$	**Force:** $F_g = \dfrac{Gm_1m_2}{r^2}$
Field Strength: $E = \dfrac{F_e}{q}$	**Field Strength:** $g = \dfrac{F_g}{m}$
Field Strength: $E = \dfrac{kq}{r^2}$	**Field Strength:** $g = \dfrac{Gm}{r^2}$
Constant: k=8.99×10⁹ N·m²/C²	**Constant:** G=6.67×10⁻¹¹ N·m²/kg²
Charge Units: Coulombs	**Mass Units:** kilograms

What is the big difference between electrostatics and gravity? The gravitational force can only attract, while the electrostatic force can both attract and repel. Notice again that both the electric field strength and the gravitational field strength follow the inverse-square law relationship. Field strength is inversely related to the square of the distance.

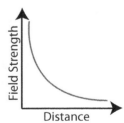

Electric Potential Difference*

When an object was lifted against the force of gravity by applying a force over a distance, work was done to give that object gravitational potential energy. The same concept applies to electric fields. If you move a charge against an electric field, you must apply a force for some distance, therefore you do work and give it electrical potential energy. The work done per unit charge in moving a charge between two points in an electric field is known as the **electric potential difference**, (V). The units of electric potential are volts, where a volt is equal to 1 Joule per Coulomb. Therefore, if you do 1 Joule of work in moving a charge of 1 Coulomb in an electric field, the electric potential difference between those points would be 1 volt. This is described mathematically by:

$$V = \frac{W}{q}$$

V in this formula is potential difference (in volts), W is work or electrical energy (in Joules), and q is your charge (in Coulombs). Let's take a look at some sample problems.

11.33 Q: A potential difference of 10 volts exists between two points, A and B, within an electric field. What is the magnitude of charge that requires 2.0×10^{-2} joules of work to move it from A to B?

11.33 A: $V = \dfrac{W}{q}$

$$q = \frac{W}{V} = \frac{2 \times 10^{-2} J}{10V} = 2 \times 10^{-3} C$$

11.34 Q: How much electrical energy is required to move a 4.00-micro-coulomb charge through a potential difference of 36.0 volts?

(A) 9.00×10^6 J
(B) 144 J
(C) 1.44×10^{-4} J
(D) 1.11×10^{-7} J

11.34 A: (C) $V = \dfrac{W}{q}$

$$W = qV = (4 \times 10^{-6} C)(36V) = 1.44 \times 10^{-4} J$$

11.35 Q: If 1.0 joule of work is required to move 1.0 coulomb of charge between two points in an electric field, the potential difference between the two points is

(A) 1.0×10^0 V

(B) 9.0×10^9 V

(C) 6.3×10^{18} V

(D) 1.6×10^{-19} V

11.35 A: (A) $V = \dfrac{W}{q} = \dfrac{1J}{1C} = 1V = 1.0 \times 10^0 V$

11.36 Q: Five coulombs of charge are moved between two points in an electric field where the potential difference between the points is 12 volts. How much work is required?

(A) 5 J

(B) 12 J

(C) 60 J

(D) 300 J

11.36 A: (C) $V = \dfrac{W}{q} \rightarrow W = qV = (5C)(12V) = 60J$

11.37 Q: In an electric field, 0.90 joules of work is required to bring 0.45 coulombs of charge from point A to point B. What is the electric potential difference between points A and B?

(A) 5.0 V

(B) 2.0 V

(C) 0.50 V

(D) 0.41 V

11.37 A: (B) $V = \dfrac{W}{q} = \dfrac{0.90J}{0.45C} = 2V$

When dealing with electrostatics, often times the amount of electric energy or work done on a charge is a very small portion of a Joule. Dealing with such small numbers is cumbersome, so physicists devised an alternate unit for electrical energy and work that can be more convenient than the Joule. This unit, known as the electronvolt (eV), is the amount of work done in moving an elementary charge through a potential difference of 1V. One electron-volt, therefore, is equivalent to one volt multiplied by one elementary charge (in Coulombs): 1 eV = 1.6×10^{-19} Joules.

11.38 Q: A proton is moved through a potential difference of 10 volts in an electric field. How much work, in electronvolts, was required to move this charge?

11.38 A: $V = \dfrac{W}{q}$

$W = qV = (1e)(10V) = 10eV$

Parallel Plates*

If you know the potential difference between two parallel plates, you can easily calculate the electric field strength between the plates. As long as you're not near the edge of the plates, the electric field is constant between the plates and the magnitude of its strength is given by the equation:

$$\left| E \right| = \frac{V}{d}$$

You'll note that with the potential difference V in volts, and the distance between the plates in meters, units for the electric field strength are volts per meter [V/m]. Previously, the units for electric field strength were given as newtons per Coulomb [N/C]. It is easy to show these are equivalent:

$$\frac{N}{C} = \frac{N \bullet m}{C \bullet m} = \frac{J}{C \bullet m} = \frac{J\!/\!C}{m} = \frac{V}{m}$$

11.39 Q: The magnitude of the electric field strength between two oppositely charged parallel metal plates is 2.0×10^3 newtons per coulomb. Point P is located midway between the plates.

(A) Sketch at least five electric field lines to represent the field between the two oppositely charged plates.

(B) An electron is located at point P between the plates. Calculate the magnitude of the force exerted on the electron by the electric field.

(C) If the plates are separated by a distance of 1 mm, determine the potential difference across the plates.

11.39 A: (A)

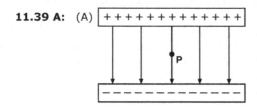

     Chapter 11: Electrostatics

(B) $E = \dfrac{F_e}{q}$

$$F_e = qE = (1.6 \times 10^{-19}\,C)(2 \times 10^3\,{}^N\!/_C) = 3.2 \times 10^{-16}\,N$$

(C) $\left|E\right| = \dfrac{V}{d} \rightarrow V = Ed = (2 \times 10^3\,{}^N\!/_C)(0.001m) = 2V$

11.40 Q: A moving electron is deflected by two oppositely charged parallel plates, as shown in the diagram below.

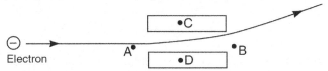

The electric field between the plates is directed from

(A) A to B

(B) B to A

(C) C to D

(D) D to C

11.40 A: (C) C to D because the electron feels a force opposite the direction of the electric field due to its negative charge.

11.41 Q: An electron is located in the electric field between two parallel metal plates as shown in the diagram below.

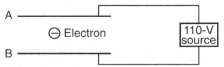

If the electron is attracted to plate A, then plate A is charged

(A) positively, and the electric field runs from plate A to plate B

(B) positively, and the electric field runs from plate B to plate A

(C) negatively, and the electric field runs from plate A to plate B

(D) negatively, and the electric field runs from plate B to plate A

11.41 A: (A) positively, and the electric field runs from plate A to plate B

11.42 Q: An electron placed between oppositely charged parallel plates moves toward plate A, as represented in the diagram below.

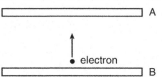

What is the direction of the electric field between the plates?

(A) toward plate A

(B) toward plate B

(C) into the page

(D) out of the page

11.42 A: (B) toward plate B.

11.43 Q: The diagram below represents two electrons, e_1 and e_2, located between two oppositely charged parallel plates.

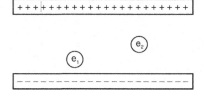

Compare the magnitude of the force exerted by the electric field on e_1 to the magnitude of the force exerted by the electric field on e_2.

11.43 A: The forces are the same because the electric field between two parallel plates is constant.

Capacitors*

Parallel conducting plates separated by an insulator can be used to store electrical charge. These devices come in a variety of sizes, and are known as parallel plate **capacitors**. The amount of charge a capacitor can store on a single plate for a given amount of potential difference across the plates is known as the device's capacitance, given in coulombs per volt, also known as a **farad** (F). A farad is a very large amount of capacitance; therefore most capacitors have values in the microfarad, nanofarad, and even pico-farad ranges.

$$C = \frac{q}{V}$$

11.44 Q: A capacitor stores 3 microcoulombs of charge with a potential difference of 1.5 volts across the plates. What is the capacitance?

11.44 A: $C = \dfrac{q}{V} = \dfrac{3 \times 10^{-6}\,C}{1.5V} = 2 \times 10^{-6}\,F$

11.45 Q: How much charge sits on the top plate of a 200 nF capacitor when charged to a potential difference of 6 volts?

11.45 A: $C = \dfrac{q}{V} \rightarrow q = CV = (200 \times 10^{-9}\,F)(6V) = 1.2 \times 10^{-6}\,C$

The amount of charge a capacitor can hold is determined by its geometry as well as the insulating material between the plates. The capacitance is directly related to the area of the plates, and inversely related to the separation between the plates, as shown in the formula below. The permittivity of an insulator (ε) describes the insulator's resistance to the creation of an electric field, and is equal to 8.85×10⁻¹² farads per meter for an air-gap capacitor.

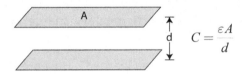

$$C = \frac{\varepsilon A}{d}$$

11.46 Q: Find the capacitance of two parallel plates of length 1 millimeter and width 2 millimeters if they are separated by 3 micrometers of air.

11.46 A: $C = \dfrac{\varepsilon A}{d} = \dfrac{(8.85 \times 10^{-12}\,{}^{F}\!/_{m})(.001m \times .002m)}{3 \times 10^{-6}\,m} = 5.9 \times 10^{-12}\,F$

11.47 Q: Find the distance between the plates of a 5 nanofarad air-gap capacitor with a plate area of 0.06 m².

11.47 A: $C = \dfrac{\varepsilon A}{d} \rightarrow d = \dfrac{\varepsilon A}{C} \rightarrow$

$d = \dfrac{(8.85 \times 10^{-12}\,{}^{F}\!/_{m})(0.06m^{2})}{5 \times 10^{-9}\,F} = 1.1 \times 10^{-4}\,m$

Equipotential Lines*

Much like looking at a topographic map which shows you lines of equal altitude, or equal gravitational potential energy, you can make a map of the electric field and connect points of equal electrical potential. These lines, known as **equipotential lines**, always cross electrical field lines at right angles, and show positions in space with constant electrical potential. If you move a charged particle in space, and it always stays on an equipotential line, no work will be done.

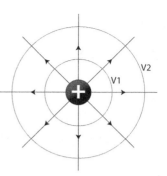

Test Your Understanding

1. Have charges smaller than an elementary charge ever been isolated? Explain.

2. What conservation laws have you learned? How are they all alike? How are they different?

3. How can you determine if an object is charged?

4. Consider the relationship of charge and electric potential difference. Compare and contrast this relationship to the relationship between mass and change in height above the Earth's surface.

5. Assume each end of a battery is attached to opposite plates of a capacitor. Why does each plate obtain the same magnitude of charge?

Chapter 12: Current Electricity

"And God said, 'Let there be light' and there was light, but the electricity board said He would have to wait until Thursday to be connected."

— Spike Milligan

Objectives

1. Define and calculate electric current.
2. Define and calculate resistance using Ohm's law.
3. Explain the factors and calculate the resistance of a conductor.
4. Identify the path and direction of current flow in a circuit.
5. Draw and interpret schematic diagrams of circuits.
6. Effectively use and analyze voltmeters and ammeters.
7. Solve series and parallel circuit problems using VIRP tables.
8. Calculate equivalent resistances for resistors in both series and parallel configurations.
9. Calculate power and energy used in electric circuits.

Electric Current

Electric current is the flow of charge, much like water currents are the flow of water molecules. Water molecules tend to flow from areas of high gravitational potential energy to low gravitational potential energy. Electric currents flow from high electric potential to low electric potential. The greater the difference between the high and low potential, the more current that flows!

In a majority of electric currents, the moving charges are negative electrons. However, due to historical reasons dating back to Ben Franklin, we say that conventional current flows in the direction positive charges would move. Although inconvenient, it's fairly easy to keep straight if you just remember that the actual moving charges, the electrons, flow in a direction opposite that of the electric current. With this in mind, you can state that positive current flows from high potential to low potential, even though the charge carriers (electrons) actually flow from low to high potential.

Electric current (I) is measured in amperes (A), or amps, and can be calculated by finding the total amount of charge (Δq), in coulombs, which passes a specific point in a given time (t). Electric current can therefore be calculated as:

$$I = \frac{\Delta q}{t}$$

12.01 Q: A charge of 30 Coulombs passes through a 24-ohm resistor in 6.0 seconds. What is the current through the resistor?

12.01 A: $I = \frac{\Delta q}{t} = \frac{30C}{6s} = 5A$

12.02 Q: Charge flowing at the rate of 2.50×10^{16} elementary charges per second is equivalent to a current of

(A) 2.50×10^{13} A

(B) 6.25×10^5 A

(C) 4.00×10^{-3} A

(D) 2.50×10^{-3} A

12.02 A: (C) $I = \frac{\Delta q}{t} = \frac{(2.50 \times 10^{16})(1.6 \times 10^{-19}C)}{1s} = 4 \times 10^{-3}A$

12.03 Q: The current through a lightbulb is 2.0 amperes. How many coulombs of electric charge pass through the lightbulb in one minute?

(A) 60 C

(B) 2.0 C

(C) 120 C

(D) 240 C

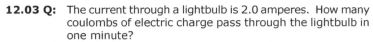

12.03 A: (C) $I = \dfrac{\Delta q}{t}$

$$\Delta q = It = (2A)(60s) = 120C$$

12.04 Q: A 1.5-volt, AAA cell supplies 750 milliamperes of current through a flashlight bulb for 5 minutes, while a 1.5-volt, C cell supplies 750 milliamperes of current through the same flashlight bulb for 20 minutes. Compared to the total charge transferred by the AAA cell through the bulb, the total charge transferred by the C cell through the bulb is

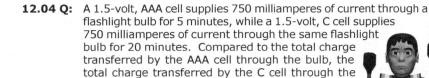

(A) half as great

(B) twice as great

(C) the same

(D) four times as great

12.04 A: (D) If Δq=It, and both cells supply 0.750A but the C cell supplies the same current for four times as long, it must supply four times the total charge compared to the AAA cell.

12.05 Q: The current traveling from the cathode to the screen in a television picture tube is 5.0×10⁻⁵ amperes. How many electrons strike the screen in 5.0 seconds?

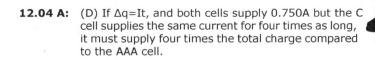

(A) 3.1×10²⁴

(B) 6.3×10¹⁸

(C) 1.6×10¹⁵

(D) 1.0×10⁵

12.05 A: (C) $I = \dfrac{\Delta q}{t}$

$$\Delta q = It = (5\times10^{-5}A)(5s) = 2.5\times10^{-4}C$$

$$2.5\times10^{-4}C \bullet \frac{1\ \text{electron}}{1.6\times10^{-19}C} = 1.6\times10^{15}\ \text{electrons}$$

Resistance

Electrical charges can move easily in some materials (conductors) and less freely in others (insulators). A material's ability to conduct electric charge is known as its **conductivity**. Good conductors have high conductivities. The conductivity of a material depends on:

1. Density of free charges available to move
2. Mobility of those free charges

In similar fashion, material's ability to resist the movement of electric charge is known as its **resistivity**, symbolized with the Greek letter rho (ρ). Resistivity is measured in ohm-meters, which is represented by the Greek letter omega multiplied by meters (Ω•m). Both conductivity and resistivity are properties of a material.

When an object is created out of a material, the material's tendency to conduct electricity, or conductance, depends on the material's conductivity as well as the material's shape. For example, a hollow cylindrical pipe has a higher conductivity of water than a cylindrical pipe filled with cotton. However, the shape of the pipe also plays a role. A very thick but short pipe can conduct lots of water, yet a very narrow, very long pipe can't conduct as much water. Both geometry of the object and the object's composition influence its conductance.

Focusing on an object's ability to resist the flow of electrical charge, objects made of high resistivity materials tend to impede electrical current flow and have a high resistance. Further, materials shaped into long, thin objects also increase an object's electrical resistance. Finally, objects typically exhibit higher resistivities at higher temperatures. You must take all of these factors into account together to describe an object's resistance to the flow of electrical charge. Resistance is a functional property of an object that describes the object's ability to impede the flow of charge through it. Units of resistance are ohms (Ω).

For any given temperature, you can calculate an object's electrical resistance, in ohms, using the following formula.

$$R = \frac{\rho L}{A}$$

Resistivities at 20°C	
Material	**Resistivity (Ω•m)**
Aluminum	2.82×10^{-8}
Copper	1.72×10^{-8}
Gold	2.44×10^{-8}
Nichrome	$150. \times 10^{-8}$
Silver	1.59×10^{-8}
Tungsten	5.60×10^{-8}

In this formula, R is the resistance of the object, in ohms (Ω), rho (ρ) is the resistivity of the material the object is made out of, in ohm•meters (Ω•m), L is the length of the object, in meters, and A is the cross-sectional area of the object, in meters squared. Note that a table of material resistivities for a constant temperature is given to you on the previous page as well.

12.06 Q: A 3.50-meter length of wire with a cross-sectional area of 3.14×10^{-6} m² at 20° Celsius has a resistance of 0.0625 Ω. Determine the resistivity of the wire and the material it is made out of.

12.06 A: $R = \dfrac{\rho L}{A}$

$$\rho = \frac{RA}{L} = \frac{(.0625\Omega)(3.14 \times 10^{-6}\, m^2)}{3.5m} = 5.6 \times 10^{-8}\, \Omega \bullet m$$

Material must be tungsten.

12.07 Q: The electrical resistance of a metallic conductor is inversely proportional to its

(A) temperature
(B) length
(C) cross-sectional area
(D) resistivity

12.07 A: (C) straight from the formula.

12.08 Q: At 20°C, four conducting wires made of different materials have the same length and the same diameter. Which wire has the least resistance?

(A) aluminum
(B) gold
(C) nichrome
(D) tungsten

12.08 A: (B) gold because it has the lowest resistivity.

12.09 Q: A length of copper wire and a 1.00-meter-long silver wire have the same cross-sectional area and resistance at 20°C. Calculate the length of the copper wire.

12.09 A:
$$R = \left(\frac{\rho L}{A}\right)_{copper} = \left(\frac{\rho L}{A}\right)_{silver}$$

$$R = \frac{\rho_{copper} L_{copper}}{A} = \frac{\rho_{silver} L_{silver}}{A}$$

$$L_{copper} = \frac{\rho_{silver} L_{silver}}{\rho_{copper}} = \frac{(1.59 \times 10^{-8}\,\Omega m)(1m)}{1.72 \times 10^{-8}\,\Omega m}$$

$$L_{copper} = 0.924m$$

12.10 Q: A 10-meter length of copper wire is at 20°C. The radius of the wire is 1.0×10⁻³ meter.

Cross Section of Copper Wire

r = 1.0 × 10⁻³ m

(A) Determine the cross-sectional area of the wire.

(B) Calculate the resistance of the wire.

12.10 A: (A) $Area_{circle} = \pi r^2 = \pi(1.0 \times 10^{-3}\,m)^2 = 3.14 \times 10^{-6}\,m^2$

(B) $R = \frac{\rho L}{A} = \frac{(1.72 \times 10^{-8}\,\Omega m)(10m)}{3.14 \times 10^{-6}\,m^2} = 5.5 \times 10^{-2}\,\Omega$

Ohm's Law

If resistance opposes current flow, and potential difference promotes current flow, it only makes sense that these quantities must somehow be related. George Ohm studied and quantified these relationships for conductors and resistors in a famous formula now known as **Ohm's Law**:

$$R = \frac{V}{I}$$

Ohm's Law may make more qualitative sense if it is rearranged slightly:

$$I = \frac{V}{R}$$

Now it's easy to see that the current flowing through a conductor or resistor (in amps) is equal to the potential difference across the object (in volts) divided by the resistance of the object (in ohms). If you want a large current to flow, you require a large potential difference (such as a large battery), and/or a very small resistance.

12.11 Q: The current in a wire is 24 amperes when connected to a 1.5 volt battery. Find the resistance of the wire.

12.11 A: $R = \frac{V}{I} = \frac{1.5V}{24A} = 0.0625\Omega$

12.12 Q: In a simple electric circuit, a 24-ohm resistor is connected across a 6-volt battery. What is the current in the circuit?
(A) 1.0 A
(B) 0.25 A
(C) 140 A
(D) 4.0 A

12.12 A: (B) $I = \frac{V}{R} = \frac{6V}{24\Omega} = 0.25A$

12.13 Q: What is the current in a 100-ohm resistor connected to a 0.40-volt source of potential difference?
(A) 250 mA
(B) 40 mA
(C) 2.5 mA
(D) 4.0 mA

12.13 A: (D) $I = \frac{V}{R} = \frac{0.40V}{100\Omega} = 0.004A = 4mA$

12.14 Q: A constant potential difference is applied across a variable resistor held at constant temperature. Which graph best represents the relationship between the resistance of the variable resistor and the current through it?

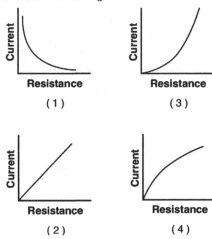

(1)　　　　　　(3)

(2)　　　　　　(4)

12.14 A: (1) due to Ohm's Law (I=V/R).

Note: Ohm's Law isn't truly a law of physics -- not all materials obey this relationship. It is, however, a very useful empirical relationship that accurately describes key electrical characteristics of conductors and resistors. One way to test if a material is ohmic (if it follows Ohm's Law) is to graph the voltage vs. current flow through the material. If the material obeys Ohm's Law, you get a linear relationship, where the slope of the line is equal to the material's resistance.

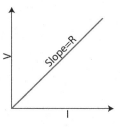

12.15 Q: The graph below represents the relationship between the potential difference (V) across a resistor and the current (I) through the resistor.

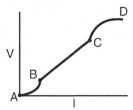

Through which entire interval does the resistor obey Ohm's law?

(A) AB

(B) BC

(C) CD

(D) AD

12.15 A: (B) BC because the graph is linear in this interval.

Electrical Circuits

An **electrical circuit** is a closed loop path through which current can flow. An electrical circuit can be made up of almost any materials (including humans if they're not careful), but practically speaking, circuits are typically comprised of electrical devices such as wires, batteries, resistors, and switches. Conventional current will flow through a complete closed loop (closed circuit) from high potential to low potential. Therefore, electrons actually flow in the opposite direction, from low potential to high potential. If the path isn't a closed loop (and is, instead, an open circuit), no current will flow.

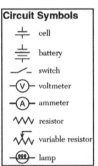

Electric circuits, which are three-dimensional constructs, are typically represented in two dimensions using diagrams known as **circuit schematics**. These schematics are simplified, standardized representations in which common circuit elements are represented with specific symbols, and wires connecting the elements in the circuit are represented by lines. Basic circuit schematic symbols are shown in the diagram at left.

In order for current to flow through a circuit, you must have a source of potential difference. Typical sources of potential difference are voltaic cells, batteries (which are just two or more cells connected together), and power (voltage) supplies. Voltaic cells are often referred to as batteries in common terminology. In drawing a cell or battery on a circuit schematic, remember that the longer side of the symbol is the positive terminal.

Electric circuits must form a complete conducting path in order for current to flow. In the example circuit shown below left, the circuit is incomplete because the switch is open, therefore no current will flow and the lamp will not light. In the circuit below right, however, the switch is closed, creating a closed loop path. Current will flow and the lamp will light up.

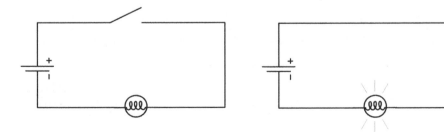

Note that in the picture at right, conventional current will flow from positive to negative, creating a clockwise current path in the circuit. The actual electrons in the wire, however, are flowing in the opposite direction, or
counter-clockwise.

Energy & Power

Just like mechanical power is the rate at which mechanical energy is expended,
electrical power is the rate at which electrical energy is expended. When
you do work on something you change its energy. Further, electrical work
or energy is equal to charge times potential difference. Therefore, you can
combine these to write the equation for electrical power as:

$$P = \frac{W}{t} = \frac{qV}{t}$$

The amount of charge moving past a point per given unit of time is current,
therefore you can continue the derivation as follows:

$$P = \left(\frac{q}{t}\right)V = IV$$

So electrical power expended in a circuit is the electrical current multiplied by
potential difference (voltage). Using Ohm's Law, you can expand this even
further to provide several different methods for calculating electrical power
dissipated by a resistor:

$$P = VI = I^2R = \frac{V^2}{R}$$

Of course, conservation of energy still applies, so the energy used in the resistor is converted into heat (in most cases) and light, or it can be used to
do work. Let's put this knowledge to use in a practical application.

12.16 Q: A 110-volt toaster oven draws a current of 6
amps on its highest setting as it converts
electrical energy into thermal energy. What
is the toaster's maximum power rating?

12.16 A: $P = VI = (110V)(6A) = 660W$

12.17 Q: An electric iron operating at 120 volts draws 10 amperes of current. How much heat energy is delivered by the iron in 30 seconds?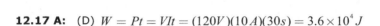

    (A) 3.0×10^2 J

    (B) 1.2×10^3 J

    (C) 3.6×10^3 J

    (D) 3.6×10^4 J

12.17 A: (D) $W = Pt = VIt = (120V)(10A)(30s) = 3.6 \times 10^4 J$

12.18 Q: One watt is equivalent to one

    (A) N·m

    (B) N/m

    (C) J·s

    (D) J/s

12.18 A: (D) J/s, since Power is W/t, and the unit of work is the joule, and the unit of time is the second.

12.19 Q: A potential drop of 50 volts is measured across a 250-ohm resistor. What is the power developed in the resistor?

    (A) 0.20 W

    (B) 5.0 W

    (C) 10 W

    (D) 50 W

12.19 A: (C) $P = \dfrac{V^2}{R} = \dfrac{(50V)^2}{250\Omega} = 10W$

12.20 Q: What is the minimum information needed to determine the power dissipated in a resistor of unknown value?

    (A) potential difference across the resistor, only

    (B) current through the resistor, only

    (C) current and potential difference, only

    (D) current, potential difference, and time of operation

12.20 A: (C) current and potential difference, only (P=VI).

Voltmeters

Voltmeters are tools used to measure the potential difference between two points in a circuit. The voltmeter is connected in parallel with the element to be measured, meaning an alternate current path around the element to be measured and through the voltmeter is created. You have connected a voltmeter correctly if you can remove the voltmeter from the circuit without breaking the circuit. In the diagram at right, a voltmeter is connected to correctly measure the potential difference across the lamp. Voltmeters have very high resistance so as to minimize the current flow through the voltmeter and the voltmeter's impact on the circuit.

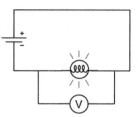

Ammeters

Ammeters are tools used to measure the current in a circuit. The ammeter is connected in series with the circuit, so that the current to be measured flows directly through the ammeter. The circuit must be broken to correctly insert an ammeter. Ammeters have very low resistance to minimize the potential drop through the ammeter and the ammeter's impact on the circuit, so inserting an ammeter into a circuit in parallel can result in extremely high currents and may destroy the ammeter. In the diagram at right, an ammeter is connected correctly to measure the current flowing through the circuit.

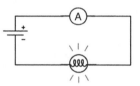

12.21 Q: In the electric circuit diagram, possible locations of an ammeter and voltmeter are indicated by circles 1, 2, 3, and 4. Where should an ammeter be located to correctly measure the total current and where should a voltmeter be located to correctly measure the total voltage?

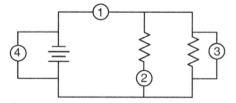

(A) ammeter at 1 and voltmeter at 4

(B) ammeter at 2 and voltmeter at 3

(C) ammeter at 3 and voltmeter at 4

(D) ammeter at 1 and voltmeter at 2

12.21 A: (A) To measure the total current, the ammeter must be placed at position 1, as all the current in the circuit must pass through this wire, and ammeters are always connected in series. To measure the total voltage in the circuit, the voltmeter could be placed at either position 3 or position 4. Voltmeters are always placed in parallel with the circuit element being analyzed, and positions 3 and 4 are equivalent because they are connected with wires (and potential is always the same anywhere in an ideal wire).

12.22 Q: Which circuit diagram below correctly shows the connection of ammeter A and voltmeter V to measure the current through and potential difference across resistor R?

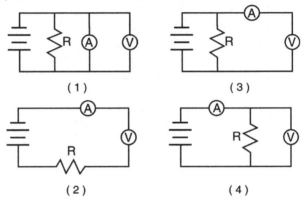

12.22 A: (4) shows an ammeter in series and a voltmeter in parallel with the resistor.

12.23 Q: A student uses a voltmeter to measure the potential difference across a resistor. To obtain a correct reading, the student must connect the voltmeter

(A) in parallel with the resistor

(B) in series with the resistor

(C) before connecting the other circuit components

(D) after connecting the other circuit components

12.23 A: (A) in parallel with the resistor.

12.24 Q: Which statement about ammeters and voltmeters is correct?

(A) The internal resistance of both meters should be low.

(B) Both meters should have a negligible effect on the circuit being measured.

(C) The potential drop across both meters should be made as large as possible.

(D) The scale range on both meters must be the same.

12.24 A: (B) Both meters should have a negligible effect on the circuit being measured.

12.25 Q: Compared to the resistance of the circuit being measured, the internal resistance of a voltmeter is designed to be very high so that the meter will draw

(A) no current from the circuit

(B) little current from the circuit

(C) most of the current from the circuit

(D) all the current from the circuit

12.25 A: (B) the voltmeter should draw as little current as possible from the circuit to minimize its effect on the circuit, but it does require some small amount of current to operate.

Series Circuits

Developing an understanding of circuits is the first step in learning about the modern-day electronic devices that dominate what is becoming known as the "Information Age." A basic circuit type, the **series circuit**, is a circuit in which there is only a single current path. Kirchhoff's Laws provide the tools in order to analyze any type of circuit.

Kirchhoff's Current Law (KCL), named after German physicist Gustav Kirchhoff, states that the sum of all current entering any point in a circuit has to equal the sum of all current leaving any point in a circuit. More simply, this is another way of looking at the law of conservation of charge.

Kirchhoff's Voltage Law (KVL) states that the sum of all the potential drops in any closed loop of a circuit has to equal zero. More simply, KVL is a method of applying the law of conservation of energy to a circuit.

12.26 Q: A 3.0-ohm resistor and a 6.0-ohm resistor are connected in series in an operating electric circuit. If the current through the 3.0-ohm resistor is 4.0 amperes, what is the potential difference across the 6.0-ohm resistor?

Chapter 12: Current Electricity

12.26 A: First, draw a picture of the situation. If 4 amps of current is flowing through the 3-ohm resistor, then 4 amps of current must be flowing through the 6-ohm resistor according to Kirchhoff's

Current Law. Since you know the current and the resistance, you can calculate the voltage drop across the 6-ohm resistor using Ohm's Law: V=IR=(4A)(6Ω)=24V.

12.27 Q: The diagram below represents currents in a segment of an electric circuit.

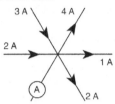

What is the reading of ammeter A?

(A) 1 A

(B) 2 A

(C) 3 A

(D) 4 A

12.27 A: (B) Since five amps plus the unknown current are coming in to the junction, and seven amps are leaving, KCL says that the total current in must equal the total current out, therefore the unknown current must be two amps in to the junction.

Let's take a look at a sample circuit, consisting of three 2000-ohm (2 kilo-ohm) resistors:

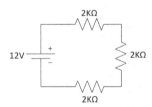

There is only a single current path in the circuit, which travels through all three resistors. Instead of using three separate 2KΩ (2000Ω) resistors, you could replace the three resistors with one single resistor having an equivalent resistance. To find the equivalent resistance of any number of series resistors, just add up their individual resistances:

$$R_{eq} = R_1 + R_2 + R_3 + \ldots$$
$$R_{eq} = 2000\Omega + 2000\Omega + 2000\Omega$$
$$R_{eq} = 6000\Omega = 6K\Omega$$

Note that because there is only a single current path, the same current must flow through each of the resistors.

A simple and straightforward method for analyzing circuits involves creating a VIRP table for each circuit you encounter. Combining your knowledge of Ohm's Law, Kirchhoff's Current Law, Kirchhoff's Voltage Law, and equivalent resistance, you can use this table to solve for the details of any circuit.

A VIRP table describes the potential drop (V-voltage), current flow (I-current), resistance (R) and power dissipated (P-power) for each element in your circuit, as well as for the circuit as a whole. Let's use the circuit with the three 2000-ohm resistors as an example to demonstrate how a VIRP table is used. To create the VIRP table, first list the circuit elements, and total, in the rows of the table, then make columns for V, I, R, and P:

VIRP Table

	V	I	R	P
R_1				
R_2				
R_3				
Total				

Next, fill in the information in the table that is known. For example, you know the total voltage in the circuit (12V) provided by the battery, and you know the values for resistance for each of the individual resistors:

	V	I	R	P
R_1			2000Ω	
R_2			2000Ω	
R_3			2000Ω	
Total	12V			

Once the initial information has been filled in, you can also calculate the total resistance, or equivalent resistance, of the entire circuit. In this case, the equivalent resistance is 6000 ohms.

	V	I	R	P
R_1			2000Ω	
R_2			2000Ω	
R_3			2000Ω	
Total	12V		6000Ω	

Looking at the bottom (total) row of the table, both the voltage drop (V) and the resistance (R) are known. Using these two items, the total current flow in the circuit can be calculated using Ohm's Law.

$$I = \frac{V}{R} = \frac{12V}{6000\Omega} = 0.002\,A$$

The total power dissipated can also be calculated using any of the formulas for electrical power.

$$P = \frac{V^2}{R} = \frac{(12V)^2}{6000\Omega} = 0.024W$$

More information can now be completed in the VIRP table:

	V	I	R	P
R_1			2000Ω	
R_2			2000Ω	
R_3			2000Ω	
Total	12V	0.002A	6000Ω	0.024W

Because this is a series circuit, the total current has to be the same as the current through each individual element, so you can fill in the current through each of the individual resistors:

	V	I	R	P
R_1		0.002A	2000Ω	
R_2		0.002A	2000Ω	
R_3		0.002A	2000Ω	
Total	12V	0.002A	6000Ω	0.024W

Finally, for each element in the circuit, you now know the current flow and the resistance. Using this information, Ohm's Law can be applied to obtain the voltage drop (V=IR) across each resistor. Power can also be found for each element using P=I²R to complete the table.

	V	I	R	P
R₁	4V	0.002A	2000Ω	**0.008W**
R₂	4V	0.002A	2000Ω	**0.008W**
R₃	4V	0.002A	2000Ω	**0.008W**
Total	12V	0.002A	6000Ω	0.024W

So what does this table really tell you now that it's completely filled out? You know the potential drop across each resistor (4V), the current through each resistor (2 mA), and the power dissipated by each resistor (8 mW). In addition, you know the total potential drop for the entire circuit is 12V, and the entire circuit dissipated 24 mW of power. Note that for a series circuit, the sum of the individual voltage drops across each element equal the total potential difference in the circuit, the current is the same throughout the circuit, and the resistances and power dissipated values add up to the total resistance and total power dissipated. These are summarized for you in the formulas below:

$$I = I_1 = I_2 = I_3 = ...$$
$$V = V_1 + V_2 + V_3 + ...$$
$$R_{eq} = R_1 + R_2 + R_3 + ...$$

12.28 Q: A 2.0-ohm resistor and a 4.0-ohm resistor are connected in series with a 12-volt battery. If the current through the 2.0-ohm resistor is 2.0 amperes, the current through the 4.0-ohm resistor is

(A) 1.0 A

(B) 2.0 A

(C) 3.0 A

(D) 4.0 A

12.28 A: (B) The current through a series circuit is the same everywhere, therefore the correct answer must be 2.0 amperes.

12.29 Q: In the circuit represented by the diagram, what is the reading of voltmeter V?

(A) 20 V

(B) 2.0 V

(C) 30 V

(D) 40 V

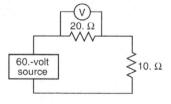

12.29 A: (D) Voltmeter reads potential difference across R_1 which is 40 V.

	V	I	R	P
R_1	40V	2A	20Ω	80W
R_2	20V	2A	10Ω	40W
Total	60V	2A	30Ω	120W

12.30 Q: In the circuit diagram below, two 4.0-ohm resistors are connected to a 16-volt battery as shown.

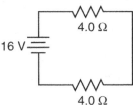

The rate at which electrical energy is expended in this circuit is

(A) 8.0 W

(B) 16 W

(C) 32 W

(D) 64 W

12.30 A: (C) 32W. Rate at which energy is expended is known as power.

	V	I	R	P
R_1	8V	2A	4Ω	16W
R_2	8V	2A	4Ω	16W
Total	16V	2A	8Ω	32W

12.31 Q: A 50-ohm resistor, an unknown resistor R, a 120-volt source, and an ammeter are connected in a complete circuit. The ammeter reads 0.50 ampere.

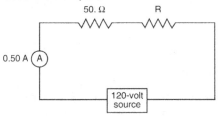

(A) Calculate the equivalent resistance of the circuit.

(B) Determine the resistance of resistor R.

(C) Calculate the power dissipated by resistor R.

12.31 A: (A) R_{eq} = 240Ω   (B) R= 190Ω   (C) $P_{50Ω\ resistor}$ = 47.5W

	V	I	R	P
R_1	25V	0.50A	50Ω	12.5W
R_2	95V	0.50A	190Ω	47.5W
Total	120V	0.50A	240Ω	60W

12.32 Q: What must be inserted between points A and B
to establish a steady electric current in the incomplete circuit represented in the diagram?

(A) switch

(B) voltmeter

(C) magnetic field source

(D) source of potential difference

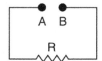

12.32 A: (D) a source of potential difference is required to drive current.

Parallel Circuits

Another basic circuit type is the **parallel circuit**, in which there is more
than one current path. To analyze resistors in a series circuit, you found an
equivalent resistance. You'll follow the same strategy in analyzing resistors
in parallel.

Let's examine a circuit made of the same components used in the exploration
of series circuits, but now connect the components so as to provide multiple
current paths, creating a parallel circuit.

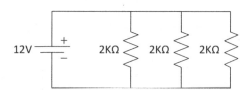

Notice that in this circuit, electricity can follow one of three different paths
through each of the resistors. In many ways, this is similar to a river branch-
ing into three different smaller rivers. Each resistor, then, causes a potential drop (analogous to a waterfall), then the three rivers recombine before
heading back to the battery, which you can think of like a pump, raising the
river to a higher potential before sending it back on its looping path. Or you
can think of it as students rushing out of a classroom. The more doors in
the room, the less resistance there is to exiting!

You can find the equivalent resistance of resistors in parallel using the formula:

$$\frac{1}{R_{eq}} = \frac{1}{R_1} + \frac{1}{R_2} + \frac{1}{R_3} + \dots$$

Take care in using this equation, as it's easy to make errors in performing your calculations. For only two resistors, this simplifies to:

$$R_{eq} = \frac{R_1 R_2}{R_1 + R_2}$$

Let's find the equivalent resistance for the sample circuit.

$$\frac{1}{R_{eq}} = \frac{1}{R_1} + \frac{1}{R_2} + \frac{1}{R_3} + \dots$$

$$\frac{1}{R_{eq}} = \frac{1}{2000\Omega} + \frac{1}{2000\Omega} + \frac{1}{2000\Omega}$$

$$\frac{1}{R_{eq}} = 0.0015 \, \frac{1}{\Omega}$$

$$R_{eq} = \frac{1}{0.0015 \, \frac{1}{\Omega}} = 667\Omega$$

A VIRP table can again be used to analyze the circuit, beginning by filling in what is known directly from the circuit diagram.

VIRP Table

	V	I	R	P
R$_1$			2000Ω	
R$_2$			2000Ω	
R$_3$			2000Ω	
Total	12V			

You can also see from the circuit diagram that the potential drop across each resistor must be 12V, since the ends of each resistor are held at a 12-volt difference by the battery

	V	I	R	P
R$_1$	12V		2000Ω	
R$_2$	12V		2000Ω	
R$_3$	12V		2000Ω	
Total	12V			

Next, you can use Ohm's Law to fill in the current through each of the individual resistors since you know the voltage drop across each resistor (I=V/R) to find I=0.006A.

	V	I	R	P
R₁	12V	**0.006A**	2000Ω	
R₂	12V	**0.006A**	2000Ω	
R₃	12V	**0.006A**	2000Ω	
Total	12V			

Using Kirchhoff's Current Law, you can see that if 0.006A flows through each of the resistors, these currents all come together to form a total current of 0.018A.

	V	I	R	P
R₁	12V	0.006A	2000Ω	
R₂	12V	0.006A	2000Ω	
R₃	12V	0.006A	2000Ω	
Total	12V	**0.018A**		

Because each of the three resistors has the same resistance, it only makes sense that the current would be split evenly between them. You can confirm the earlier calculation of equivalent resistance by calculating the total resistance of the circuit using Ohm's Law: R=V/I=(12V/0.018A)=667Ω.

	V	I	R	P
R₁	12V	0.006A	2000Ω	
R₂	12V	0.006A	2000Ω	
R₃	12V	0.006A	2000Ω	
Total	12V	0.018A	**667Ω**	

Finally, you can complete the VIRP table using any of the three applicable equations for power dissipation to find:

	V	I	R	P
R₁	12V	0.006A	2000Ω	**0.072W**
R₂	12V	0.006A	2000Ω	**0.072W**
R₃	12V	0.006A	2000Ω	**0.072W**
Total	12V	0.018A	667Ω	**0.216W**

Note that for resistors in parallel, the equivalent resistance is always less than the resistance of any of the individual resistors. The potential difference across each of the resistors in parallel is the same, and the current through each of the resistors adds up to the total current. This is summarized for you in the following table:

$$I = I_1 + I_2 + I_3 + ...$$
$$V = V_1 = V_2 = V_3 = ...$$
$$\frac{1}{R_{eq}} = \frac{1}{R_1} + \frac{1}{R_2} + \frac{1}{R_3} + ...$$

12.33 Q: A 15-ohm resistor, R_1, and a 30-ohm resistor, R_2, are to be connected in parallel between points A and B in a circuit containing a 90-volt battery.

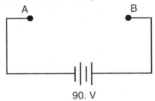

90. V

(A) Complete the diagram to show the two resistors connected in parallel between points A and B.

(B) Determine the potential difference across resistor R_1.

(C) Calculate the current in resistor R_1.

12.33 A: (A)

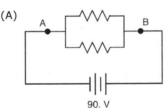

90. V

(B) Potential difference across R_1 is 90V.

(C) Current through resistor R_1 is 6A.

	V	I	R	P
R_1	90V	6A	15Ω	540W
R_2	90V	3A	30Ω	270W
Total	90V	9A	10Ω	810W

12.34 Q: Draw a diagram of an operating circuit that includes: a battery as a source of potential difference, two resistors in parallel with each other, and an ammeter that reads the total current in the circuit.

12.34 A:

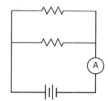

12.35 Q: Three identical lamps are connected in parallel with each other. If the resistance of each lamp is X ohms, what is the equivalent resistance of this parallel combination?

(A) X Ω

(B) X/3 Ω

(C) 3X Ω

(D) 3/X Ω

12.35 A: (B) X/3 Ω

$$\frac{1}{R_{eq}} = \frac{1}{R_1} + \frac{1}{R_2} + \frac{1}{R_3} + \dots$$

$$\frac{1}{R_{eq}} = \frac{1}{X} + \frac{1}{X} + \frac{1}{X}$$

$$\frac{1}{R_{eq}} = \frac{3}{X}$$

$$R_{eq} = \frac{X}{3}$$

12.36 Q: Three resistors, 4 ohms, 6 ohms, and 8 ohms, are connected in parallel in an electric circuit. The equivalent resistance of the circuit is

(A) less than 4 Ω

(B) between 4 Ω and 8 Ω

(C) between 10 Ω and 18 Ω

(D) 18 Ω

12.36 A: (A) the equivalent resistance of resistors in parallel is always less than the value of the smallest resistor.

12.37 Q: A 3-ohm resistor, an unknown resistor, R, and two ammeters, A_1 and A_2, are connected as shown with a 12-volt source. Ammeter A_2 reads a current of 5 amperes.

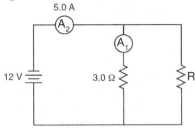

(A) Determine the equivalent resistance of the circuit.

(B) Calculate the current measured by ammeter A_1.

(C) Calculate the resistance of the unknown resistor, R.

12.37 A: (A) 2.4Ω   (B) 4A   (C) 12Ω

	V	I	R	P
R_1	12V	4A	3Ω	48W
R_2	12V	1A	12Ω	12W
Total	12V	5A	2.4Ω	60W

12.38 Q: The diagram below represents an electric circuit consisting of four resistors and a 12-volt battery.

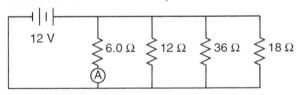

(A) What is the current measured by ammeter A?

(B) What is the equivalent resistance of this circuit?

(C) How much power is dissipated in the 36-ohm resistor?

12.38 A: (A) 2A   (B) 3Ω   (C) 4W

	V	I	R	P
R_1	12V	2A	6Ω	24W
R_2	12V	1A	12Ω	12W
R_3	12V	0.33A	36Ω	4W
R_4	12V	0.67A	18Ω	8W
Total	12V	4A	3Ω	48W

12.39 Q: A 20-ohm resistor and a 30-ohm resistor are connected in parallel to a 12-volt battery as shown. An ammeter is connected as shown.

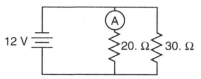

(A) What is the equivalent resistance of the circuit?

(B) What is the current reading of the ammeter?

(C) What is the power of the 30-ohm resistor?

12.39 A: (A) 12Ω   (B) 0.6A   (C) 4.8W

	V	I	R	P
R₁	12V	0.6A	20Ω	7.2W
R₂	12V	0.4A	30Ω	4.8W
Total	12V	1A	12Ω	12W

12.40 Q: In the circuit diagram shown below, ammeter A_1 reads 10 amperes.

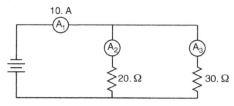

What is the reading of ammeter A_2?

(A) 6 A

(B) 10 A

(C) 20 A

(D) 4 A

12.40 A: (A) 6 A

	V	I	R	P
R₁	120V	6A	20Ω	720W
R₂	120V	4A	30Ω	480W
Total	120V	10A	12Ω	1200W

Combination Series-Parallel Circuits

A circuit doesn't have to be completely serial or parallel. In fact, most circuits actually have elements of both types. Analyzing these circuits can be accomplished using the fundamentals you learned in analyzing series and parallel circuits separately and applying them in a logical sequence.

First, look for portions of the circuit that have parallel elements. Since the voltage across the parallel elements must be the same, replace the parallel resistors with an equivalent single resistor in series and draw a new schematic. Now you can analyze your equivalent series circuit with a VIRP table. Once your table is complete, work back to your original circuit using KCL and KVL until you know the current, voltage, and resistance of each individual element in your circuit.

12.41 Q: Find the current through R_2 in the circuit below.

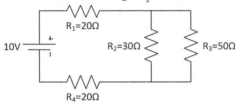

12.41 A: First, find the equivalent resistance for R_2 and R_3 in parallel.

$$R_{eq_{23}} = \frac{R_2 R_3}{R_2 + R_3} = \frac{(30\Omega)(50\Omega)}{30\Omega + 50\Omega} = 19\Omega$$

Next, re-draw the circuit schematic as an equivalent series circuit.

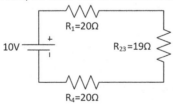

Now, you can use your VIRP table to analyze the circuit.

	V	I	R	P
R_1	3.39V	0.169A	20Ω	0.57W
R_{23}	3.22V	0.169A	19Ω	0.54W
R_4	3.39V	0.169A	20Ω	0.57W
Total	10V	0.169A	59Ω	1.69W

The voltage drop across R_2 and R_3 must therefore be 3.22 volts. From here, you can apply Ohm's Law to find the current through R_2:

$$I_2 = \frac{V_2}{R_2} = \frac{3.22V}{30\Omega} = 0.107A$$

	V	I	R	P
R_1	3.39V	0.169A	20Ω	0.57W
R_2	3.22V	0.107A	30Ω	0.34W
R_3	3.22V	0.062A	50Ω	0.20W
R_4	3.39V	0.169A	20Ω	0.57W
Total	10V	0.169A	59Ω	1.69W

12.42 Q: Consider the following four DC circuits.

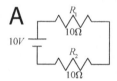

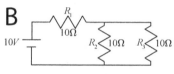

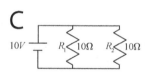

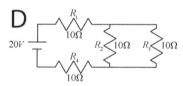

For each circuit, rank the following quantities from highest to lowest in terms of:

I) Current through R_2.

II) Power dissipated by R_1.

III) Equivalent resistance of the entire circuit.

IV) Potential drop across R_2.

12.42 A: I) C, A, D, B

II) C, D, B, A

III) D, A, B, C

IV) C, A, D, B

VIRP Tables for Each Circuit Below:

Circuit A	V	I	R	P
R_1	5V	0.5A	10Ω	2.5W
R_2	5V	0.5A	10Ω	2.5W
Total	10V	0.5A	20Ω	5W

Circuit B	V	I	R	P
R_1	6.7V	0.67A	10Ω	4.5W
R_2	3.3V	0.33A	10Ω	1.1W
R_3	3.3V	0.33A	10Ω	1.1W
Total	10V	0.67A	15Ω	6.7W

Circuit C	V	I	R	P
R_1	10V	1A	10Ω	10W
R_2	10V	1A	10Ω	10W
Total	10V	2A	5Ω	20W

Circuit D	V	I	R	P
R_1	8V	0.8A	10Ω	6.4W
R_2	4V	0.4A	10Ω	1.6W
R_3	4V	0.4A	10Ω	1.6W
R_4	8V	0.8A	10Ω	6.4W
Total	20V	0.8A	25Ω	16W

Test Your Understanding

1. If all you have are 20-ohm resistors, how could you make a 5-ohm resistor? How could you make a 30-ohm resistor?

2. Explain how Kirchhoff's Current Law (the junction rule) is a restatement of the law of conservation of electrical charge.

3. Explain how Kirchhoff's Voltage Law (the loop rule) is a restatement of the law of conservation of energy.

4. What happens to the power usage in a room if you replace a 100-watt lightbulb with two 60-watt lightbulbs in parallel?

5. What is the resistance of an ideal ammeter? What is the resistance of an ideal voltmeter? What would happen if you switched those meters in a circuit?

6. Graph the current through, voltage across, and power dissipated by a variable resistor attached to a constant-voltage battery as the resistor's resistance increases linearly with time.

Appendix A: AP-Style Problems

Kinematics

1. A ball is thrown vertically upward from the ground. Which pair of graphs best describes the motion of the ball as a function of time while it is in the air? Neglect air resistance.

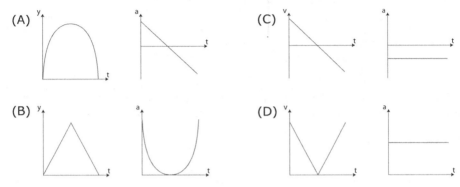

Answer: C

The velocity of the ball starts at a high initial value (assuming the upward direction is positive), then slows down at a constant rate until it reaches it highest point, at which point its velocity reaches an instantaneous value of zero. The ball then accelerates downward, increasing its speed at a constant rate until it returns to the same point in space and same speed it began its journey with. Throughout the entire time interval, the acceleration remains a constant negative value (a=-g=-9.81 m/s² on the surface of the Earth).

---

2. Two children on the playground, Bobby and Sandy, travel down slides of identical height h but different shapes as shown at right. The slides are frictionless. Assuming they start down the slides at the same time with zero initial velocity, which of the following statements is true?

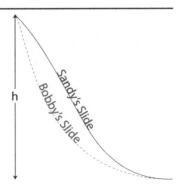

(A) Bobby reaches the bottom first with the same average velocity as Sandy.

(B) Bobby reaches the bottom first with a larger average acceleration than Sandy.

(C) Bobby reaches the bottom first with the same average acceleration as Sandy.

(D) They reach the bottom at the same time with the same average acceleration.

Answer: B

Both children begin with gravitational potential energy mgh at the top of the slide, which is completely transferred to kinetic energy at the end of the slide. Bobby's potential energy is transferred more quickly, however, therefore he attains a higher average velocity and beats Sandy to the end of the slide. Average acceleration is the change in velocity divided by the time interval. Each child has the same change in velocity, but Bobby observes this change over a shorter period of time, resulting in a larger average acceleration.

---

3. The following projectiles are launched on horizontal ground with the same initial speed. If two or three situations have the same answer, put the letters in the same blank space. Air resistance is negligible.

(a) Rank the situations from least to greatest in terms of time the projectile is in the air.

Least 1_____ 2_____ 3_____ 4_____ Greatest

Or, all of the projectiles are the same _____

(b) Rank the situations from least to greatest in terms of the horizontal distance the projectile travels while in the air.

Least 1_____ 2_____ 3_____ 4_____ Greatest

Or, all of the projectiles are the same _____

(c) Rank the situations from least to greatest in terms of the vertical distance the projectile travels while in the air.

Least 1_____ 2_____ 3_____ 4_____ Greatest

Or, all of the projectiles are the same _____

(d) Rank the situations from least to greatest in terms of the acceleration of the projectile while it is in the air.

Least 1_____ 2_____ 3_____ 4_____ Greatest

Or, all of the projectiles are the same _____

Answers:

(a) A, B, C, D -- the more vertical the vector the more time in the air since all initial velocities are the same.
(b) D, CA, B -- The 45° projectile has the greatest horizontal range.
(c) A, B, C, D -- D has the largest vertical component of initial velocity.
(d) All are the same acceleration.

---

4) You are asked to experimentally determine the acceleration of a skier traveling down a snow-covered hill of uniform slope as accurately as possible. Which combination of equipment and equation would be most useful in your endeavor?

equipment	equation
(A) tape measure, stopwatch	$x = x_0 + v_{x0}t + \frac{1}{2}a_x t^2$
(B) photo gates, stopwatch	$v_x^2 = v_{x0}^2 + 2a_x(x - x_0)$
(C) radar gun, tape measure	$v_x = v_{x0} + a_x t$
(D) photo gates, radar gun	$\overline{v}_x = \dfrac{v_{0x} + v_x}{2}$

Answer: A

Measure the time it takes the skier to go a set distance using the stopwatch, and use the tape measure to determine the distance. You may then use the equation $x = x_0 + v_{x0}t + \frac{1}{2}a_x t^2$ to solve for the acceleration, recognizing that the initial velocity and position of the skier are zero.

---

5. An eagle flies at constant velocity horizontally across the sky, carrying a turtle in its talons. The eagle releases the turtle while in flight. From the eagle's perspective, the turtle falls vertically with speed v_1. From an observer on the ground's perspective, at a particular instant the turtle falls at an angle with speed v_2. What is the speed of the eagle with respect to an observer on the ground?

(A) $v_1 + v_2$

(B) $v_1 - v_2$

(C) $\sqrt{v_1^2 - v_2^2}$

(D) $\sqrt{v_2^2 - v_1^2}$

Answer: D

Call the velocity of the turtle with respect to the eagle v_{TE}, also known as v_1.
Call the velocity of the turtle with respect to the ground v_{TG}, also known as v_2.
You are asked to find the velocity of the eagle with respect to the ground, v_{EG}.

Analyzing the right triangle, you can use the Pythagorean Theorem to solve for the magnitude of v_{EG}.

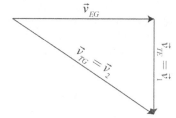

$$\vec{v}_{TG} = \vec{v}_{EG} + \vec{v}_{TE}$$

$$\vec{v}_{TG} = \vec{v}_{EG} + \vec{v}_{TE} \xrightarrow[\vec{v}_{TE}=\vec{v}_1]{\vec{v}_{TG}=\vec{v}_2} \vec{v}_2 = \vec{v}_{EG} + \vec{v}_1 \rightarrow$$

$$v_2^2 = v_{EG}^2 + v_1^2 \rightarrow v_{EG} = \sqrt{v_2^2 - v_1^2}$$

---

6. A car travels through a rainstorm at constant speed v_C as shown in the diagram at right. Rain is falling vertically at a constant speed v_R with respect to the ground. If the back windshield of the car, highlighted in the diagram, is set at an angle of θ with the vertical, what is the maximum speed the car can travel and still have rain hit the back windshield?

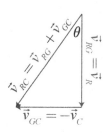

back windshield

(A) $v_R \cos\theta$ (B) $v_R \tan\theta$

(C) $v_R \sin\theta$ (D) $v_R(1-\sin\theta)$

Answer: (B) $v_R \tan\theta$

In order for the rain to just hit the windshield, the angle of the velocity vector for the rain with respect to the car must match the angle of the back windshield. The velocity of the rain with respect to the car (v_{RC}) can be found as the vector sum of the velocity of the rain with respect to the ground (v_R) and the velocity of the grouncd with respect to the car ($-v_C$) as shown in the diagram at right. From this diagram, it is a straightforward application of trigonometry to find the speed of the car at which this condition occurs.

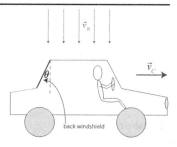

$$\tan\theta = \frac{v_{GC}}{v_{RG}} = \frac{v_C}{v_R} \rightarrow v_C = v_R \tan\theta$$

For questions 7 and 8, use the information given below.

An object slides one meter down a frictionless ramp of constant slope as shown below right (not to scale). A student measures the time it takes for the object to travel various displacements using a stopwatch. Three consecutive trials are measured, and the data is recorded as shown below.

Displacement (m)	Avg. Time (s)
0	0
0.2	0.68
0.4	0.98
0.6	1.18
0.8	1.38
1	1.52

7. Determine the acceleration of the object.

Answer: 0.855 m/s²

There are several methods of arriving at the answer, including, but not limited to, taking the slope of the v-t graph (calculating velocity for the various time intervals) or the use of the kinematic equations.

8. Determine the angle θ of the ramp.

Answer: 5°

After finding the acceleration of the object, recognize that the acceleration of the object is a=gsin(θ). Solving for theta, then, gives:

$$a = g\sin\theta \rightarrow \theta = \sin^{-1}\left(\frac{a}{g}\right) = \sin^{-1}\left(\frac{0.855\,{}^{m}\!/_{s^2}}{9.81\,{}^{m}\!/_{s^2}}\right) = 5°$$

9. A pirate captain in her ship spies her first mate in a dinghy five kilometers away. The pirate captain sails her ship toward the dinghy at a rate of eight kilometers per hour. The first mate rows his dinghy toward the pirate ship at a rate of two kilometers per hour. When the captain initially spies the first mate at a distance of five kilometers, her parrot, Polly, begins flying back and forth between the two at a rate of 40 kilometers per hour. How far does Polly fly in total if she continues her back-and-forth journey until the pirate ship meets the dinghy?

Answer: 20 kilometers

The ship and dinghy approach each other at a combined 10 km per hour, therefore it takes them 0.5 hours to meet. During this entire time period, the parrot flies at 40 km/hr, therefore the total distance traveled by the parrot is 20 kilometers.

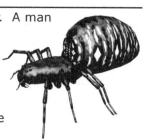

---

10. A train travels east toward Chicago at 80 km/hr. A man on the train runs from the front of the train toward the rear of the train at 10 km/hr. As he runs, he carries a plate of fruit with him. He notices a giant spider on the plate and throws the plate away from him (toward the rear of the train) at 20 km/hr. The startled spider jumps toward the man at 5 km/hr. The instant after the spider jumps toward the man, how fast is the spider approaching Chicago?

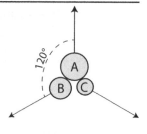

Answer: 55 km/hr

The spider's velocity with respect to the ground is the vector sum of the four given velocities, 80 km/hr east, -10 km/hr east, -20 km/hr east, and 5 km/hr east, for a resultant of 55 km/hr east.

---

11. Three penguins are arranged in the center of a patch of frictionless ice floating across the ocean with a velocity of 2 m/s west. The mass of penguin A is 38 kg, the mass of penguin B is 30 kg, and the mass of penguin C is 23 kg. At time t=0, the penguins push off each other, each with a force of 20 newtons, such that they all slide away from the center of the floating ice patch at an angle of 120° from each other as shown in the diagram at right. Describe the motion of the center of mass of the three-penguin system at time t=3s.

Answer: The center of mass continues in its current state of motion, traveling west at 2 m/s.

Because there are no external forces on the system, the velocity of the center of mass of the system cannot be changed.

12. A fisherman in a small fishing boat at rest in a lake hooks a giant log floating in the lake 30 meters away. The fisherman reels the log in. During this process, the boat moves 12 meters in the direction of the log. If the mass of the boat and fisherman is 400 kg, what is the mass of the log? Assume frictionless.

Answer: m_{log} = 267 kg

Because there are no external forces on the system, and the center of mass is initially at rest, the center of mass must remain at rest. Set x=0 as the center of mass of the fisherman / log system:

$$x_{cm} = \frac{x_{fisherman}m_{fisherman} + x_{log}m_{log}}{m_{fisherman} + m_{log}} = 0 \rightarrow x_{fisherman}m_{fisherman} + x_{log}m_{log} = 0 \rightarrow$$

$$m_{log} = \frac{-x_{fisherman}m_{fisherman}}{x_{log}} = \frac{(12m)(400kg)}{(18m)} = 267kg$$

---

13. Two balls are launched off the edge a cliff of height h with an initial velocity v_0. The red ball is launched horizontally. The green ball is launched at an angle of θ above the horizontal. Neglect air resistance.

(a) Derive an expression for the time the red ball is in the air.
(b) Derive an expression for the horizontal distance traveled by the red ball while it is in the air.
(c) Derive an expression for the time the green ball is in the air.
(d) Derive an expression for the horizontal distance traveled by the green while while it is in the air.
(e) If the initial launch velocity v_0 of the balls is 100 m/s, the green ball is launched at an angle θ=30°, and the balls land 600 meters apart from each other, what is the height of the cliff? (Note: calculator use strongly encouraged for this step).

Answers:

(a) $t_{red} = \sqrt{\dfrac{2h}{g}}$

(b) $\Delta x_{red} = v_0\sqrt{\dfrac{2h}{g}}$

(c) $t_{green} = \dfrac{v_0 \sin\theta + \sqrt{v_0^2 \sin^2\theta + 2gh}}{g}$

(d) $\Delta x_{green} = (v_0\cos\theta)\left(\dfrac{v_0\sin\theta + \sqrt{v_0^2 \sin^2\theta + 2gh}}{g}\right)$

(e) H=81m

Dynamics

1. A mixed martial artist kicks his opponent in the nose with a force of 200 newtons. Identify the action-reaction force pairs in this interchange.

(A) foot applies 200 newton force to nose; nose applies a smaller force to foot because foot has a larger mass.

(B) foot applies 200 newton force to nose; nose applies a smaller force to foot because it compresses.

(C) foot applies 200 newton force to nose; nose applies a larger force to foot due to conservation of momentum.

(D) foot applies 200 newton force to nose; nose applies an equal force to the foot.

Answer: (D) foot applies 200 newton force to nose; nose applies an equal force to the foot. This is a basic application of Newton's 3rd Law.

---

2. Joanne exerts a force on a basketball as she throws the basketball to the east. Which of the following is always true?

(A) Joanne accelerates to the west.

(B) Joanne feels no net force because she and the basketball are initially the same object.

(C) The basketball pushes Joanne to the west.

(D) The magnitude of the force on the basketball is greater than the magnitude of the force on Joanne.

Answer: (C) The basketball pushes Joanne to the west.

Newton's 3rd Law of Motion states that if Joanne applies a force on the basketball to the east, the basketball must apply a force back on Joanne in opposite direction, to the west.

Appendix A: AP-Style Problems

3. A book and a feather are pushed off the edge of a cliff simultaneously. The book reaches the bottom of the cliff before the feather. Correct statements about the book include which of the following? Select two answers.

(A) The book has a greater mass than the feather and experiences less air resistance.

(B) The book has a greater mass than the feather and experiences a greater net force.

(C) The book has a greater cross-sectional area than the feather and experiences less air resistance.

(D) The book has a greater cross-sectional area than the feather and experiences more air resistance.

Answers: B & D

The force of air resistance on the book is greater than the force of air resistance on the feather due to the larger cross-sectional area of the book (the book moves through more air). However, the book also experiences a larger downward force due to gravity due to its larger mass. Even though the force of air resistance is greater for the book, the proportion of the force of air resistance to the force of gravity on the book is smaller than that for the feather. The book has a larger net force on it, and also a larger proportion of force to mass, resulting in a larger downward acceleration.

4. A force F is applied perpendicular to the top of a box of mass m sitting on an incline of angle θ. What is the magnitude of F such that the normal force of the incline on the box is equal to the weight of the box?

(A) $mg\cos\theta$

(B) $mg(1-\cos\theta)$

(C) $mg(1-\sin\theta)$

(D) $mg(1+\sin\theta)$

Answer: (B) $mg(1-\cos\theta)$

Begin with a diagram, free body diagram, and pseudo-free body diagram.

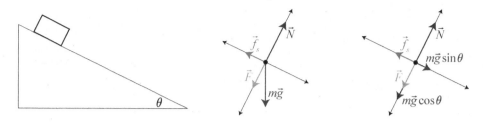

Writing a Newton's 2nd Law equation in the y-direction provides the appropriate relationship to solve for the applied force.

$$N - F - mg\cos\theta = 0 \rightarrow F = N - mg\cos\theta \xrightarrow{N=mg} F = mg - mg\cos\theta = mg(1-\cos\theta)$$

5. In 1654 in Magdeburg, Germany, scientist Otto von Guericke demonstrated the concept of atmospheric pressure by placing two sealed iron hemispheres together and using a vacuum pump to create a partial vacuum inside the spheres. He then attached a team of 15 horses to each of the hemispheres, and had the horses attempt to pull the spheres apart. All 30 horses were not able to separate the spheres.

Suppose von Guericke had instead attached both teams (all 30 horses) to one of the hemispheres and attached the remaining hemisphere to a tree. How would the tension in the spheres change?

(A) The tension in the spheres would be reduced by half.

(B) The tension in the spheres would remain the same.

(C) The tension in the spheres would double.

(D) The tension in the spheres would quadruple.

Answer: (C) The tension in the spheres would double.

When each team of horses is pulling on the spheres, the tension from each team is the same, with each team pulling in opposite directions. Whatever force one team pulls with, the other team must pull back with an equal magnitude force, otherwise the sphere would accelerate. This is the same as the force that would be exerted if one side of the sphere was held motionless while one team of horses pulls on a hemisphere. By attaching both teams (all 30 horses) to the same hemisphere, and fixing the opposing hemisphere in place, double the force of one team is obtained between the hemispheres.

6. Identical fireflies are placed in closed jars in three different configurations as shown below. In configuration A, three fireflies are hovering inside the jar. In configuration B, one firefly is hovering inside the jar. In configuration C, one firefly is sitting at rest on the bottom of the jar. Each jar is placed upon a scale and measured. Rank the weight of each jar according to the scale reading from heaviest to lightest. If jars have the same scale reading, rank them equally.

A   **B**   **C**

Answer: A, B=C

Whether the fireflies are in flight or sitting on the bottom of the jar, they provide the same weight to the scale (if they are flying in the jar, the force their wings provide on the air pushing down is equal to the force of the air pushing them up. This same air pushes down on the bottom of the jar by Newton's 3rd Law, making their weights equivalent whether flying or resting. Therefore, the only factor in determining the weight is the number of fireflies in the jar.

7. The system shown at right is accelerated by applying a tension T_1 to the right-most cable. Assuming the system is frictionless, the tension in the cable between the blocks, T_2, is

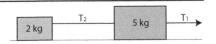

(A) $2T_1/7$

(B) $2T_1/5$

(C) $5T_1/7$

(D) $7T_1/5$

Answer: (A) $2T_1/7$

Analyzing the system as a whole, $T_1=7a$, therefore $a=T_1/7$

Looking at just the 2-kg block, $T_2=2a$. Substituting in the acceleration of the system (since both blocks must have the same acceleration), you find $T_2=2T_1/7$

---

8) Jane rides a sled down a slope of angle θ at constant speed v. Determine the coefficient of kinetic friction between the sled and the slope. Neglect air resistance.

(A) $\mu=g\sin\theta$

(B) $\mu=mg\cos\theta$

(C) $\mu=\tan\theta$

(D) $\mu=g\cos\theta$

Answer: (C) $\mu=\tan\theta$

Utilizing a free body diagram (and pseudo-free body diagram to bring the weight vector into components that line up with the axes, you find that $F_f=mg\sin\theta$ and $N=mg\cos\theta$. Putting these together to to determine the co-efficient of friction:

$$\mu = \frac{F_f}{N} = \frac{mg\sin\theta}{mg\cos\theta} = \tan\theta$$

---

9) Masses are hung on a light string attached to an ideal massless pulley as shown in the diagram at right. The total mass hanging from the left string is equal to that on the right. At time t=0, the 0.2-kg mass is moved from the left to the right side of the pulley. How far does each mass move in one second?

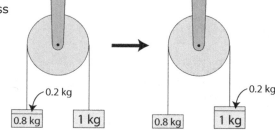

0.2 kg

0.8 kg   1 kg

0.2 kg

0.8 kg   1 kg

Answer: $\Delta y = 1$ m

First draw a free body diagram for the two masses.

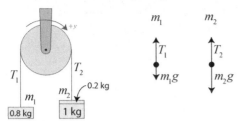

Next you can write out Newton's 2nd Law equations for the two objects, keeping in mind the y-axis we have chosen curving around the pulley.

$$T_1 - m_1g = m_1a$$
$$m_2g - T_2 = m_2a$$

Combining these equations and recognizing $T_1 = T_2$ because this is an ideal, massless pulley, you can then solve for the acceleration of the system.

$$m_2g - m_1g = (m_1 + m_2)a \rightarrow a = g\left(\frac{m_2 - m_1}{m_1 + m_2}\right) = 2\,{}^{m}\!/_{s^2}$$

With the acceleration known, you can solve for the distance traveled using kinematic equations.

$$\Delta y = v_0t + \tfrac{1}{2}a_yt^2 \rightarrow \Delta y = \tfrac{1}{2}(2\,{}^{m}\!/_{s^2})(1s)^2 = 1m$$

10. The Atwood machine shown to the right consists of a massless, frictionless pulley, and a massless string. If $m_1 = 3$ kg and $m_2 = 5$ kg, find the force of tension in the string.

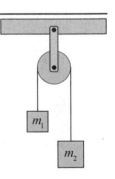

Answer: F_T=37.5 N

First step is to solve for the acceleration of the masses by using Newton's second law. The 5 kg mass is accelerating downward and the 3 kg mass is accelerating upward. Assume g = 10 m/s²

$$\sum F = ma$$
$$m_2 g - m_1 g = (m_1 + m_2)a$$
$$50N - 30N = (8kg)a$$
$$a = 2.5\,{}^m\!/_{s^2}$$

Second step is to isolate the 5 kg mass to find the tension in the string. Again use Newton's second law.

$$\sum F = ma$$
$$-F_T + m_2 g = m_2 a$$
$$-F_T + 50N = (5kg)(2.5\,{}^m\!/_{s^2})$$
$$F_T = 37.5N$$

⚖ Two masses are hung from a light string over an ideal frictionless massless pulley. The masses are shown in various scenarios in the diagram below. Rank the acceleration of the masses from greatest to least.

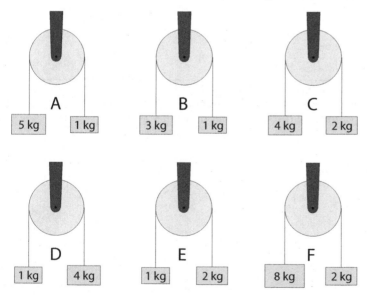

A	
5 kg	1 kg

B	
3 kg	1 kg

C	
4 kg	2 kg

D	
1 kg	4 kg

E	
1 kg	2 kg

F	
8 kg	2 kg

Answer: A > D=F > B > C=E

A simple analysis of the Atwood Machines shows that the acceleration of the system is equal to the net force divided by the sum of the masses. Summarizing in table form:

Scenario	m_1 (kg)	m_2 (kg)	F_{net} (N)	a (m/s^2)
A	5	1	40	6.67
B	3	1	20	5
C	4	2	20	3.33
D	1	4	30	6
E	1	2	10	3.33
F	8	2	60	6

12. Students build a wind-powered vehicle as shown in the diagram at right. A fan attached to the vehicle is powered by a battery located inside the vehicle. Describe the motion of the vehicle when the fan is powered on. Justify your answer in a clear, coherent, paragraph-length explanation.

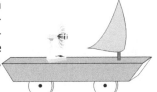

Answer: The vehicle remains at rest.

Students may approach their justification in multiple ways. An example of a strong answer could include references to all three of Newton's Laws (i.e. the vehicle will remain at rest because it is not acted upon by an outside force. You can observe this as the spinning fan imparts a force on the air molecules, which, by Newton's 3rd Law, apply an opposing force on the fan blades. The air molecules impact the sail, imparting a force on the sail, while the sail imparts an opposing force back on the air molecules. Because both the fan and the sail are part of the same object, the opposing forces on the sail and fan blades balance each other out, resulting in no net force. According to Newton's 2nd Law, if there is no net force, there is no acceleration, therefore the vehicle remains at rest.

Other possible explanations include, but are not limited to, a dynamics analysis looking at the vehicle from a systems perspective, as well as a momentum analysis.

13. The Atwood machine shown below consists of a massless, frictionless pulley, a massless string, and a massless spring. The spring has a spring constant of 200 N/m. How far will the spring stretch when the masses are released?

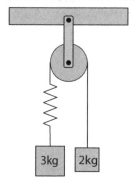

Answer: x = 0.12m

First step is to solve for the acceleration of the masses by using Newton's second law. The 3-kg mass is accelerating downward (negative direction) and the 2-kg mass is accelerating upward (positive direction). Assume g = 10 m/s².

$$\sum F = ma$$
$$-30N + 20N = (5kg)a$$
$$a = -2\,\text{m}/_{s^2}$$

Second step is to isolate the 3-kg mass to find the tension acting on the spring. Again use Newton's second law.

$$\sum F = ma$$
$$F_T - 30N = (3kg)(-2\,\text{m}/_{s^2})$$
$$F_T = 24N$$

Third step is to use Hooke's Law to solve for the stretch of the spring.

$$F = kx$$
$$24N = (200\,\text{N}/_m)x$$
$$x = 0.12m$$

14) Paisley the horse gets stuck in the mud. Her rider, Linda, ties a rope around Paisley and pulls with a force of 500 newtons, but isn't strong enough to get Paisley out of the mud.

In a flash of inspiration, Linda thinks back to her physics classes and ties one end of the rope to Paisley, and the other end to a nearby fence post. She then applies the same force of 500 newtons to the middle of the rope at an angle of θ=6°, just barely freeing Paisley from the mud, as shown in the diagram at right.

Determine the minimum amount of tension the rope must support.

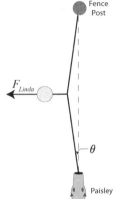

Fence
Post

F_{Linda}

θ

Paisley

Answer: 2400 N

A free body diagram showing all the forces on the rope is quite helpful in this situation. Note that the tension in the portion of rope connected to the fence post must be the same as the tension in the portion of rope connected to Paisley.

$$F_{net_x} = 2F_T \sin\theta - F_{Linda} = 0 \rightarrow F_T = \frac{F_{Linda}}{2\sin\theta} = \frac{500N}{2\sin(6°)} = 2400N$$

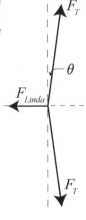

F_T

θ

F_{Linda}

F_T

15. Curves can be banked at just the right angle that vehicles traveling a specific speed can stay on the road with no friction required. Given a radius for the curve and a specific velocity, at what angle should the bank of the curve be built?

Answer: $\theta = \tan^{-1}\left(\dfrac{v^2}{gr}\right)$

First draw a free body diagram for the car on the banked curve, then a pseudo-FBD, showing all forces lining up parallel to the axes.

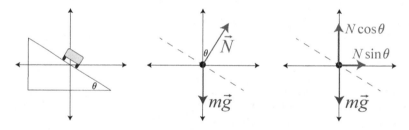

Then, write out Newton's 2nd Law equations for both the x- and y-directions.

$$F_{net_x} = F_{net_c} = N\sin\theta = \frac{mv^2}{r}$$

$$F_{net_y} = N\cos\theta - mg = 0 \rightarrow N = \frac{mg}{\cos\theta}$$

Finally, combining the results of these two equations, solve for the angle θ.

$$N\sin\theta = \frac{mv^2}{r} \xrightarrow{N=\frac{mg}{\cos\theta}} \frac{mg\sin\theta}{\cos\theta} = \frac{mv^2}{r} \rightarrow g\tan\theta = \frac{v^2}{r} \rightarrow \theta = \tan^{-1}\left(\frac{v^2}{gr}\right)$$

16 The Atwood machine shown to the right consists of a massless, frictionless pulley, a massless string, and a massless spring. The spring has a spring constant of 100 N/m. Calculate the stretch of the spring in terms of M.

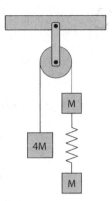

Answer: x = (0.133M) m

First Step is to solve for the acceleration of the masses by using Newton's second law. The 4m mass is accelerating downward and the other 2 masses are accelerating upward. We can put these masses on the horizontal axis and assume the masses are all accelerating in the same –x direction.

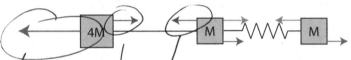

This new horizontal system can be analyzed using Newton's second law. Assume g = 10 N/kg

Analyzing forces from right to left:

$$\sum F = Ma$$
$$-4Mg + F_T - F_T + Mg + F_s - F_s + Mg = (6M)a$$
$$a = -3.3 \, {}^m\!/_{s^2}$$

Second Step is to isolate the mass on the right end to find the force acting from the spring. Again use Newton's second law. Assume g = 10 N/kg

$$\sum F = Ma$$
$$Mg - F_s = Ma$$
$$F_s = Mg - Ma = (13.3M) \, {}^m\!/_{s^2}$$

Third Step is to use Hooke's Law to solve for the stretch of the spring.

$$F_s = kx$$
$$(13.3M) \, {}^m\!/_{s^2} = (100 \, {}^N\!/_m)x$$
$$x = (0.133M)m$$

17. A student, standing on a scale in an elevator at rest, sees that the scale reads 840 N. As the elevator rises, the scale reading increases to 945 N for 2 seconds, then returns to 840 N for 10 seconds. When the elevator slows to a stop at the 10th floor, the scale reading drops to 735 N for 2 seconds while coming to a stop.

(a) Explain why the apparent weight of the student increased at the beginning of the motion.

(b) Draw the free body diagram for the student while the student is accelerating upward, then moving at a constant velocity, and finally accelerating downward at the end. Draw the length of the force vectors to show when forces are balanced or unbalanced.

(c) Sketch acceleration vs. time, velocity vs. time, and displacement vs. time graphs of the student during the elevator ride.

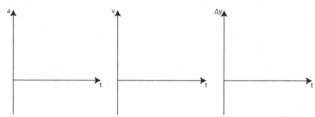

Answer:

(a) The student's mass has inertia and wants to stay at rest. In order to accelerate her mass, an unbalanced force must be applied. The reading on the scale shows not only the force of her normal weight, but has an additional unbalanced force causing her mass to accelerate.

(b)

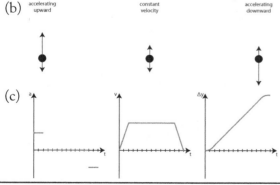

(c)

Work, Power & Energy

1. Bowling Ball A is dropped from a point halfway up a cliff. A second identical bowling ball, B, is dropped simultaneously from the top of the cliff. Comparing the bowling balls at the instant they reach the ground, which of the following are correct? Neglect air resistance.

(A) Ball A has half the kinetic energy and takes half the time to hit the ground as Ball B.

(B) Ball A has half the kinetic energy and takes one-fourth the time to hit the ground as Ball B.

(C) Ball A has half the final velocity and takes half the time to hit the ground as Ball B.

(D) Ball A has one-fourth the final velocity and takes one-fourth the time to hit the ground as Ball B.

(E) None of these are correct.

Answer: E

The final velocities of the balls are given by $v = \sqrt{2gh}$.

The final velocity of B is related to the square root of the height, therefore the final velocity of B is $\sqrt{2}$ times the final velocity of A, eliminating choices C and D. Further, the kinetic energy of Ball A is half the kinetic energy of Ball B at the instant the balls reach the ground. The time it takes for the balls to reach the ground is also related to the square root of the height, therefore the time for B to hit the ground is $\sqrt{2}$ times the time for A to hit the ground, eliminating choices A and B. E must be the correct answer.

2. A net force (Fcosθ) acts on an object in the x-direction while moving over a distance of 4 meters along the axis, depicted in the graph below.

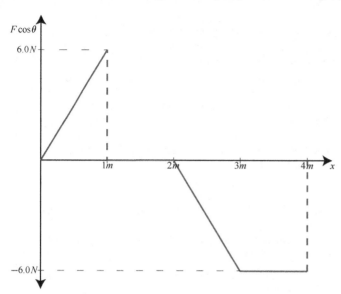

(a) Find the work done by the force in the interval from 0.0 to 1.0 m.
(b) Find the work done by the force in the interval from 1.0 to 2.0 m.
(c) Find the work done by the force in the interval from 2.0 to 4.0 m.
(d) At what position(s) is the object moving with the largest speed? Explain your answer.

Answers:

(a) 3 J
(b) 0 J
(c) -9 J
(d) The object is moving with the largest speed at the positions of 1 to 2 m. From 0 to 1 m the force is increasing the speed of the object where it reaches a maximum value. From 1 to 2 m there is no force acting on the object therefore it is moving at a constant speed. From 2 to 4 m the negative force is slowing the object down since the object is still moving in the positive x direction.

③ Bob pushes a box across a horizontal surface at a constant speed of 1 m/s. If the box has a mass of 30 kg, find the power Bob supplies given the coefficient of kinetic friction is 0.3.

Answer: 90 W

First, draw a free body diagram for the situation.

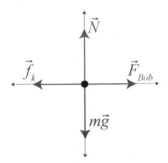

Next, recognizing the box is moving at constant speed and is therefore experiencing no net force, apply Newton's 2nd Law in the x- and y-directions to solve for Bob's applied force.

$$F_{net_X} = F_{Bob} - f_k = ma_x \xrightarrow{a_x=0} F_{Bob} = f_k = \mu_k N$$

$$F_{net_Y} = N - mg = 0 \rightarrow N = mg$$

$$F_{Bob} = \mu_k N = \mu_k mg = (0.3)(30kg)(10\,m/_{s^2}) = 90N$$

Finally, use the relationship between power, force, and velocity to determine the power supplied.

$$P = \frac{\Delta E}{\Delta t} = \frac{Fd\cos\theta}{\Delta t} \xrightarrow[\cos\theta=1]{\bar{v}=\frac{d}{\Delta t}} P = F\bar{v} = (90N)(1\,m/_s) = 90W$$

4. A roller coaster car begins at rest at height h above the ground and completes a loop along its path. In order for the car to remain on the track throughout the loop, what is the minimum value for h in terms of the radius of the loop, R? Assume frictionless.

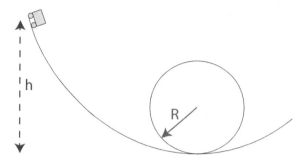

Answer: h must be greater than or equal to 5R/2.

First, use the law of conservation of energy to find the speed of the cart at the top of the loop.

$$U_i = K_f + U_f \rightarrow mgh = \tfrac{1}{2}mv_f^2 + mg(2R) \rightarrow 2gh = v_f^2 + 4gR \rightarrow v_f^2 = 2g(h - 2R)$$

Next, recognize for the cart to stay on the loop and not fall off, the centripetal acceleration must be equal to or greater than the acceleration due to gravity.

$$\frac{v^2}{R} \geq g \rightarrow v^2 \geq gR \rightarrow 2g(h - 2R) \geq gR \rightarrow (h - 2R) \geq \frac{R}{2} \rightarrow h \geq \frac{5R}{2}$$

3. A dart of mass m is accelerated horizontally through a tube of length L situated a height h above the ground by a constant force F. Upon exiting the tube, the dart travels a horizontal distance Δx before striking the ground, as depicted in the diagram below.

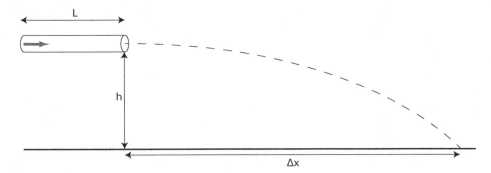

(a) Develop an expression for the velocity of the dart, v, as it leaves the tube in terms of Δx, h, and any fundamental constants.
(b) Derive an expression for the kinetic energy of the dart as it leaves the tube in terms of m, Δx, h, and any fundamental constants.
(c) Derive an expression for the work done on the dart in the tube in terms of F and L.
(d) Derive an expression for the height of the tube above the ground in terms of m, Δx, L, F, and any fundamental constants.

An experiment is then performed in which the length of the tube, L, is varied, resulting in the dart traveling various horizontal distances Δx which are recorded in the table below.

Trial	1	2	3	4	5	6
Tube Length L (m)	0.2	0.5	0.8	1.2	1.7	2.0
Horizontal Distance Δx (m)	3.5	5.5	7.0	8.6	10.2	11.1

(e) Plot a linear graph of Δx^2 as a function of L. Use the empty boxes in the data table above, as appropriate, to record the calculated values you are graphing. Label the axes as appropriate, and place numbers on both axes.
(f) From the graph, obtain the height of the tube given the mass of the dart is 20 grams and the constant force applied in the tube is 2 newtons.

Answer:

(a) $v = \dfrac{\Delta x}{\sqrt{2h/g}}$

(b) $K = \dfrac{mg\Delta x^2}{4h}$

(c) $W = FL$

(d) $h = \dfrac{mg\Delta x^2}{4FL}$

(e)

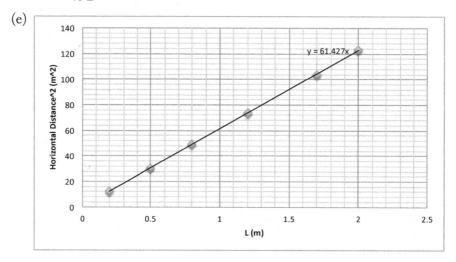

(f) $h = \dfrac{\Delta x^2}{L}\dfrac{mg}{4F} = slope\left(\dfrac{mg}{4F}\right) = 61\left[\dfrac{(0.02)(9.8)}{4(2)}\right] = 1.5m$

Linear Momentum

1. A cart full of water travels horizontally on a frictionless track with initial velocity **v**. As shown in the diagram, in the back wall of the cart there is a small opening near the bottom of the wall that allows water to stream out. Considering just the cart itself (and not the water inside it), which of the following most accurately describes the characteristics of the cart?

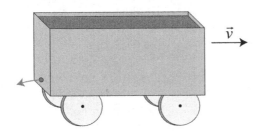

	Speed	**Kinetic Energy**
(A)	stays the same	stays the same
(B)	increases	increases
(C)	stays the same	increases
(D)	increases	stays the same

Answer: B

As the water streams out of the cart, the water is pushed out of the cart by pressure from above, exerting a reactionary force pushing the cart and its contents forward, increasing the speed of the cart. As the speed of the cart is increasing, and its mass remains the same, the kinetic energy of the cart must also increase.

Special thanks to Dan Burns, Gardner Friedlander, and Zaid Khalil for assistance with problem development.

2. A firecracker is launched with an initial velocity of 70 m/s at an angle of 73° with the horizontal. The firecracker explodes at its highest point, splitting into three equal pieces. One piece continues at its same horizontal speed, but moves vertically upward at 10 m/s immediately after the explosion. A second piece moves vertically downward at 10 m/s, but with a horizontal velocity of 30 m/s backward immediately after the explosion. Determine the speed of the remaining piece of the firecracker immediately following the explosion. Neglect air resistance.

(A) 0 m/s   (C) 70 m/s

(B) 10 m/s   (D) 90 m/s

Answer: (C) 70 m/s

You can approach this problem as either a center-of-mass problem with no external forces, or a conservation of linear momentum problem, again with no external forces. The horizontal velocity of the intact firecracker at its highest point is 20 m/s. Once the firecracker explodes, three pieces of equal mass are ejected. The vertical components of the velocity of the first and second pieces are equal in magnitude and opposite in direction, so the vertical component of velocity for the remaining piece must be zero. The horizontal component of the first piece of the firecracker has no change in horizontal velocity. Therefore, the change in horizontal velocity of the second piece must balance the change in horizontal velocity of the third piece. If the second piece is moving backward at 30 m/s (a change in velocity of -50 m/s), the third piece must experience a change in velocity of +50 m/s, for a total velocity of 70 m/s horizontally, and a final speed of 70 m/s.

---

3. A cart traveling on a smooth track with velocity v collides and sticks to an identical cart on the track, initially at rest. What is the maximum percentage of the cart's initial kinetic energy maintained as kinetic energy in the two-cart system?

(A) 25%   (C) 75%

(B) 50%   (D) 100%

Answer: (B) 50%

Consistent with the law of conservation of momentum, following the collision the maximum possible speed of the two carts combined is v/2. Kinetic Energy, however, is related to the mass and the square of the velocity. Following the collision, the kinetic energy of the combined carts doubles due to the doubling of mass, but is quartered due to the effect of $(v/2)^2$. The product of these two effects, then, is an effective reduction in the kinetic energy of the combined carts by at least 50%. The missing energy must have been lost through non-conservative means such as heat and sound.

4. A identical bullets are fired into identical wood blocks in two different positions as shown in the diagram below. In each case the wood block (with the bullet embedded) rises to a certain height before returning to the surface of the Earth. Assuming the bullets have the same initial velocities, which block will go higher, and why?

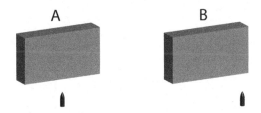

(A) Block A will go higher since all of the bullet's initial kinetic energy is transformed into gravitational potential energy (height) in scenario A, while some of the bullet's initial kinetic energy in scenario B is transformed into rotational kinetic energy, therefore B doesn't go as high.

(B) Block B will go higher since the rotational kinetic energy in situation B adds to the translational kinetic energy imparted by the bullet, while Block A doesn't receive this additional rotational kinetic energy.

(C) The blocks will reach the same maximum height due to conservation of linear momentum.

(D) The blocks will reach the same maximum height due to conservation of angular momentum.

Answer: (C) The blocks will reach the same maximum height due to conservation of linear momentum. It is true that block B will also have some amount of rotational kinetic energy, and both blocks have the same maximum gravitational potential energy. This difference in energy is accounted for in realizing this is an inelastic collisions, and the difference in energy of the blocks is due to differences in penetration of the bullets into the blocks.

Note that this experiment, as well as an explanation of results, is available from Derek Muller's Veritasium YouTube Channel, and is a highly recommended follow-up exercise to this problem: https://www.youtube.com/watch?v=vWVZ6APXM4w

5. Two small, uniform balls of identical density and size are fired from a toy gun toward a wooden block. Ball A is highly elastic and bounces backward after striking the block. Ball B is made of clay and sticks to the wooden block upon impact. Which of the following statements best describes the effects of the collision with the block?

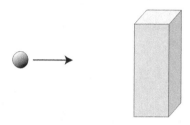

(A) Ball A transfers more momentum and more energy to the block than Ball B.

(B) Ball A transfers more momentum and less energy to the block than Ball B.

(C) Ball A transfers less momentum and more energy to the block than Ball B.

(D) Ball A transfers less momentum and less energy to the block than Ball B.

Answer: (B) Ball A transfers more momentum and less energy to the block than Ball B.

Both balls have the same initial momentum prior to striking the block. Following the collision, however, the elastic ball, Ball A, bounces backward, transferring up to twice its initial momentum to the block through the larger impulse. Ball B, however, sticks to the block, transferring its initial momentum to the block through an impulse equal to its initial momentum. Therefore, Ball A transfers more momentum to the block.

With respect to energy transfer, however, the story changes. Ball A maintains some kinetic energy as it rebounds off the block, therefore it cannot transfer as much kinetic energy to the block as Ball B, which transfers all of its kinetic energy to the block as it comes to rest. Therefore, Ball B transfers more kinetic energy to the block.

6. An 80 kg student stands on the left end of a 240-kg log which is floating in the water (which you may treat as a frictionless surface). The student and the log are both initially at rest.

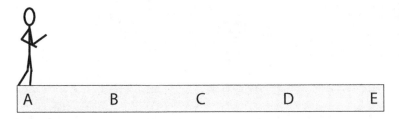

| A | B | C | D | E |

(a) Which point is the approximate center of mass of the student-log system?
(b) The student walks to the far end of the log at a constant speed of 2 m/s. As the student walks to the right, describe the motion of the log. Provide quantitative values wherever possible.
(c) What is the velocity of the center of mass of the student-log system while the student is walking?

Answer:

(a) (B) is the approximate center of mass of the system.
(b) The log moves to the left with a speed of 0.67 m/s as the student moves right.
(c) The velocity of the center of mass of the student-log system is 0 m/s while the student is walking since the initial velocity of the center of mass of the system was 0 m/s and no external forces were applied.

⑦ A 0.5-kilogram block slides at 20 m/s on a smooth frictionless surface to-ward a stationary sphere, shown below.

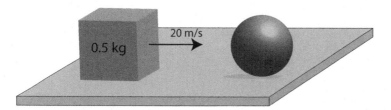

The sphere is half the volume of the block, but is eight times as dense. The block strikes the sphere at time t=0. A plot of the force exerted on the cube by the ball as a function of time is shown below.

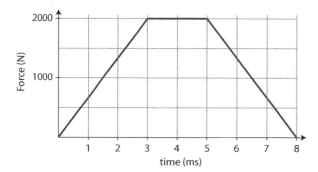

(a) What is the impulse applied to the block?
(b) What is the speed of the ball immediately following the collision?
(c) What is the velocity of the cube immediately following the collision? (State both direction and magnitude.)
(d) Is this an elastic collision? Justify your answer.

Answer:

(a) The impulse is the area under the curve which is 10 N•s.
(b) The impulse applied is equal to the change of momentum, therefore

$$J = \Delta p = m\Delta v \rightarrow \Delta v = \frac{J}{m} = \frac{10N \bullet s}{2kg} = 5 \,{}^m\!/_s$$

(c) The change in velocity of the block can also be found from the impulse-momentum theorem:

$$J = \Delta p = m\Delta v \rightarrow \Delta v = \frac{J}{m} = \frac{-10N \bullet s}{0.5kg} = -20 \,{}^m\!/_s$$

The velocity of the block after the collision is therefore 0 m/s, since it was going 20 m/s, and its change of velocity is -20 m/s. The direction is unnecessary -- the block is stationary.
(d) This is an inelastic collision since kinetic energy is not conserved. The initial kinetic energy of the block-sphere system is 100 J, and the final kinetic energy of the block-sphere system is 25 J.

8. A proton (mass=m) and a lithium nucleus (mass=7m) undergo an elastic collision as shown below.

m — v=1000 m/s →   7m

Find the velocity of the lithium nucleus following the collision.

Answer: 250 m/s

First, solve for the velocity of the proton as a function of the velocity of the lithium nucleus using conservation of linear momentum.

$$\vec{p}_b = \vec{p}_a \rightarrow 1000m = mv_p = 7mv_L \rightarrow v_p = 1000 - 7v_L$$

Next, utilize conservation of kinetic energy since this is an elastic collision to sole for the velocity of the lithium nucleus.

$$K_b = K_a \rightarrow \tfrac{1}{2}m(1000)^2 = \tfrac{1}{2}mv_p^2 + \tfrac{1}{2}(7m)v_L^2 \rightarrow 10^6 = v_p^2 + 7v_L^2 = (1000 - 7v_L)^2 + 7v_L^2 \rightarrow$$
$$v_L = 250 \, m/\!s$$

Circular Motion & Rotation

1. A 50-kg boy and a 40-kg girl sit on opposite ends of a 3-meter see-saw. How far from the girl should the fulcrum be placed in order for the boy and girl to balance on opposite ends of the see-saw?

Answer: 1.67 m

In order for the children to balance, the net torque about the fulcrum must be zero, resulting in no angular acceleration. A diagram such as the one below may be helpful in developing the mathematical relationships.

$$\vec{\tau}_{net} = I\vec{\alpha} = 0 \rightarrow (50)(g)(3-x) - (40)(g)(x) = 0 \rightarrow 500(3-x) - 400x = 0 \rightarrow 1500 - 500x - 400x = 0 \rightarrow$$
$$x = 1.67m$$

---

2. A uniform hollow tube of length L rotates vertically about its center of mass as shown. A ball is dropped into the tube at position A, and exits a short time later at position B. From the perspective of a stationary observer watching the tube rotate, the distance the ball travels is

(A) less than L

(B) greater than L

(C) equal to L

Answer: (B) greater than L

Though the displacement of the ball at B from its initial position at A is less than the length of the rod, L, the distance the ball travels is greater than L from the point of view of an external observer watching the tube rotate. Imagine the tube is transparent as you observe the path of the ball from a stationary reference point. You would observe the ball traveling a curved path from A to B, as shown at right. The length of A to the center point is one radius, and since the ball takes a curved path, it travels a distance greater than that radius. The same occurs from the center point to point B. Therefore, the ball travels a distance greater than the length of the tube from the perspective of an external observer at a stationary reverence point.

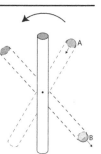

---

3. Four particles, each of mass M, move in the x-y plane with varying velocities as shown in the diagram. The velocity vectors are drawn to scale. Rank the magnitude of the angular momentum about the origin for each particle from largest to smallest.

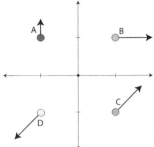

Answer: C, B, A, D

C has the highest angular momentum as it has the maximum velocity and the angle between the position vector to the mass and the velocity vector is 90 degrees. B has the next highest angular momentum as it has the maximum velocity, and the angle between the position vector to the mass and the velocity vector is 45 degrees. A has the third highest angular momentum as it has half the maximum velocity, and the angle between the position vector to the mass and the velocity vector is 45 degrees. D has the lowest angular momentum (0) about the origin since the position vector to the mass and its velocity are parallel.

4. A disc rotates clockwise about its axis as shown in the diagram. The direction of the angular momentum vector is:

(A) out of the plane of the page

(B) into the plane of the page

(C) toward the top of the page

(D) toward the bottom of the page

Answer: (B) into the plane of the page

The direction of the angular momentum vector is given by the right-hand-rule. In this case, wrap the fingers of your right-hand around the axle in the direction of the rotational velocity, and your thumb points in the direction of the angular momentum vector (into the plane of the page.)

5. A hoop with moment of inertia I=0.1 kg•m² spins about a frictionless axle with an angular velocity of 5 radians per second. At what radius from the center of the hoop should a force of 2 newtons be applied for 3 seconds in order to accelerate the hoop to an angular speed of 10 radians per second?

(A) 8.3 cm

(B) 12.5 cm

(C) 16.7 cm

(D) 25 cm

Answer: (A) 8.3 cm

A net torque will change the angular momentum of the hoop. Solving for the distance at which the force must be applied to create the appropriate torque:

$$\tau_{net} = I\alpha \rightarrow Fr = I\left(\frac{\omega_f - \omega_i}{t}\right) \rightarrow r = \frac{I}{F}\left(\frac{\omega_f - \omega_i}{t}\right) = \frac{0.1}{2}\left(\frac{10-5}{3}\right) = 0.083m = 8.3cm$$

---

6. Jean stands at the exact center of a large spinning frictionless uniform disk of mass M and radius R with moment of inertia $I=\frac{1}{2}MR^2$. As she walks from the center to the edge of the disk, the angular speed of the disk is quartered. Which of the following statements is true?

(A) Jean's mass is less than the mass of the disk.

(B) Jean's mass is equal to the mass of the disk.

(C) Jean's mass is between the mass of the disk and twice the mass of the disk.

(D) Jean's mass is more than twice the mass of the disk.

Answer: (C) Jean's mass is between the mass of the disk and twice the mass of the disk.

The initial moment of inertia of the system is approximately $I=\frac{1}{2}MR^2$. Jean's mass does not contribute significantly to this moment of inertia as she stands at the exact center of the large disk. As she walks from the center of the disk to the edge of the disk, however, she adds her moment of inertia to that of the disk, so that the total moment of inertia of the system is now $I=\frac{1}{2}MR^2 + mR^2$. Calling the initial angular speed of the disk ω and applying the law of conservation of angular momentum, find Jean's mass (m) as follows:

$$L_i = L_f \rightarrow \frac{1}{2}MR^2\omega = (\frac{1}{2}MR^2 + mR^2)\frac{\omega}{4} \rightarrow 4M = M + 2m \rightarrow m = \frac{3}{2}M$$

7. A spinning plate in a microwave with moment of inertia I rotates about its center of mass at a constant angular speed ω. When the microwave ends its cook cycle, the plate comes to rest in time Δt due to a constant frictional force F applied a distance r from the axis of rotation. What is the magnitude of the frictional force F?

(A) $F = \dfrac{I\omega}{r\Delta t}$

(C) $F = \dfrac{\omega r^2}{I\Delta t}$

(B) $F = \dfrac{Ir}{\omega\Delta t}$

(D) $F = \dfrac{Ir^2}{\omega\Delta t}$

Answer: (A) $F = \dfrac{I\omega}{r\Delta t}$

The change in angular momentum of the plate is given by the product of the net torque and the time interval over which it is applied. Therefore:

$$\Delta L = I\Delta\omega = \tau\Delta t \rightarrow I\omega = Fr\Delta t \rightarrow F = \dfrac{I\omega}{r\Delta t}$$

8. A planet orbits a sun in an elliptical orbit as shown. Which principles of physics most clearly and directly explain why the speed of the planet is the same at positions A and B? Select two answers.

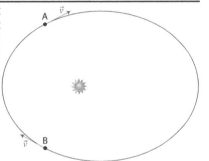

(A) Conservation of Energy

(B) Conservation of Angular Velocity

(C) Conservation of Angular Momentum

(D) Conservation of Charge

Answer: (A) Conservation of Energy and (C) Conservation of Angular Momentum

(A) Ultimately conservation of energy can lead you to this conclusion, though there are several steps analyzing kinetic and gravitational potential energy to get there.

(B) There is no such law as Conservation of Angular Velocity. This is silly.

(C) An analysis using conservation of Angular Momentum leads directly to $m_A v_A R_A \sin\theta_A = m_B v_B R_B \sin\theta_B$, and given that R, θ, and m are the same at positions A and B, you have a clear and direct path to $v_A = v_B$.

(D) Conservation of electrical charge, though true, does not help you with this problem.

9. A given force is applied to a wrench to turn a bolt of specific rotational inertia I which rotates freely about its center as shown in the following diagrams. Which of the following correctly ranks the resulting angular acceleration of the bolt?

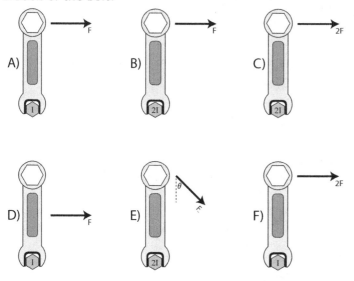

(A) $\alpha_F > \alpha_A = \alpha_C > \alpha_B = \alpha_D > \alpha_E$

(B) $\alpha_C = \alpha_F > \alpha_A = \alpha_B > \alpha_D > \alpha_E$

(C) $\alpha_A = \alpha_F > \alpha_C = \alpha_E > \alpha_B = \alpha_D$

(D) $\alpha_D = \alpha_F > \alpha_A = \alpha_C > \alpha_B > \alpha_E$

Answer: (A) $\alpha_F > \alpha_A = \alpha_C > \alpha_B = \alpha_D > \alpha_E$

Net torque is equal to the product of the angular acceleration and the rotational inertia, therefore angular acceleration is net torque divided by moment of inertia. Assume the length of the entire wrench is L.

Diagram	Torque	Rotational I	Angular Acceleration
A	FL	I	FL/I
B	FL	2I	FL/2I
C	2FL	2I	FL/I
D	FL/2	I	FL/2I
E	FLsin45°	2I	FL√2/4I
F	2FL	I	2FL/I

10. Gina races her bike across a horizontal path. Suddenly, a squirrel runs in front of her. Gina slams on both her front and rear brakes, which results in the bike flipping over the front wheel and Gina flying over the handle bars. Which of the following do NOT contribute to an explanation of why the bike flips and Gina flies over the handlebars?

(A) Gina has a tendency to continue moving at a constant velocity, so while the bike stops, Gina continues her previous motion.

(B) Conservation of angular momentum of the bike and wheels indicates that if the wheels stop spinning in one direction, the bike must spin in the opposite direction.

(C) The large negative acceleration of the bike/rider system reduces the moment of inertia of the system, increasing the system's angular acceleration and causing a rotation of the bike.

(D) The force of the applied brakes at a distance from the center of mass of the bike and rider produces a net torque on the bike, causing a rotation bringing the back wheel of the back up.

Answer: (C) The large negative acceleration of the bike/rider system reduces the moment of inertia of the system, increasing the system's angular acceleration and causing a rotation of the bike.

(A) is true, as a restatement of Newton's 1st Law of Motion. (B) is true, though the effect may be rather small if the mass of the wheels is relatively small compared to the mass of the rest of the bike. (D) is true as the force of friction provides a net torque on the bike (similar to how a motorcycle may "pop a wheelie" when accelerated quickly). (C) is completely made up, however, and is incorrect.

A 20-kg ladder of length 8 m sits against a frictionless wall at an angle of 60 degrees. The ladder just barely keeps from slipping.

(a) On the diagram below, draw and label the forces acting on the ladder.

(b) Determine the force of friction of the floor on the ladder.
(c) Determine the coefficient of friction between the ladder and the floor.

Answer:

(a)

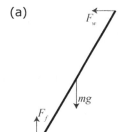

(b) Utilize Newton's 2nd Law in the x- and y-directions, as well as Newton's 2nd Law for Rotation, to determine the force of friction f.

$$F_{net_X} = f - F_w = 0 \rightarrow f = F_w$$

$$F_{net_Y} = F_f - mg = 0 \rightarrow F_f = mg = (20kg)(10\,^m/_{s^2}) = 200N$$

$$\tau_{net} = -4mg\cos 60° + 8F_w \sin 60° \rightarrow F_w = \frac{4mg\cos 60°}{8\sin 60°} = 57.7N$$

(c) Solve for the coefficient of friction: $f = \mu N \rightarrow \mu = \dfrac{f}{N} = \dfrac{57.7N}{200N} = 0.29$

A student designs an experiment in which a mass (m) attached to a one-meter-long string is wrapped around a pulley of mass M=1 kg and radius R=0.1 m and dropped. The pulley includes a sensor which measures and records its rotational velocity. The mass is dropped from rest and the final rotational velocity of the pulley as well as the time the mass is in the air is measured. Data is shown in the table below.

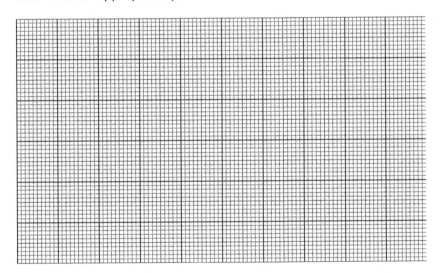

m (kg)	ω_f (rad/s)	time (s)	α (rad/s²)
0.2	24	0.84	
0.4	30	0.68	
0.6	33	0.61	
0.8	35	0.58	
1.0	36	0.55	

(a) Complete the table above by calculating the angular acceleration for each of the hanging masses.

(b) Plot the angular acceleration vs. the hanging mass. Make sure to label all axes appropriately.

(c) Using your plot, estimate the angular acceleration of the pulley for a hanging mass of 500g.

(d) Assume the sensor which measures the rotational velocity of the pulley breaks. Explain how you could accomplish this experiment without using the sensor.

(e) Estimate the time it will take for a 500g mass attached to the string to fall all the way through the pulley. Show all work.

Answer:

(a)

m (kg)	ω_f (rad/s)	time (s)	α (rad/s²)
0.2	24	0.84	28
0.4	30	0.68	44
0.6	33	0.61	54
0.8	35	0.58	60
1.0	36	0.55	65

(b)

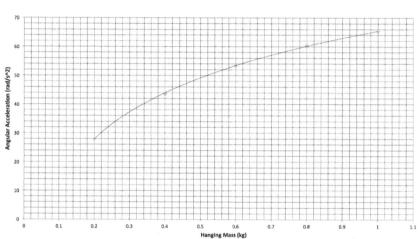

(c) $\alpha \approx 49 \, {}^{rad}\!/_{s^2}$ (read from graph)

(d) Knowing the time it takes for the hanging mass to fall one meter, you could calculate the linear acceleration of the falling mass, which would correspond to the angular acceleration of the pulley using the relationship $\alpha = a/R$.

(e) The linear acceleration can be determined from a simple transformation:

$$a = \alpha R = (49 \, {}^{rad}\!/_{s^2})(0.1m) = 4.9 \, {}^{m}\!/_{s^2}$$

Then, time can be determined using basic kinematics:

$$\Delta y = v_0 t + \tfrac{1}{2} a_y t^2 \xrightarrow{v_0 = 0} t = \sqrt{\frac{2\Delta y}{a_y}} = \sqrt{\frac{2(1m)}{4.9 \, {}^{m}\!/_{s^2}}} = 0.64s$$

13. A particle of mass m is launched with velocity v toward a uniform disk of mass M and radius R which can rotate about a point on its edge as shown. The disk is initially at rest. After the particle strikes the edge of the disk and sticks, the magnitude of the final angular velocity of the disk-particle system is given by:

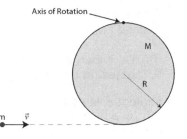

$$|\vec{\omega}| = \left(\frac{4m}{8m+3M}\right)\left(\frac{v}{R}\right)$$

Determine the moment of inertia of the disk about its axis of rotation.

Answer: $I_{disk} = \frac{3}{2}MR^2$

Treating the disk and particle as a single system, the total angular momentum of the system must be conserved in the absence of external torques, therefore the initial angular momentum of the system must equal the final angular momentum of the system.

$$L_i = L_f = I\omega_f$$

Substituting in the initial angular momentum of the particle and the final angular velocity of the system and solving for the total moment of inertia of the disk-particle system:

$$mv(2R) = I_{total}\omega_f = I_{total}\left(\frac{4m}{8m+3M}\right)\left(\frac{v}{R}\right) \rightarrow I_{total} = 4mR^2 + \frac{3}{2}MR^2$$

Finally, recognizing the total moment of inertia of the system is the sum of the moment of inertia of the particle and the moment of inertia of the disk, solve for the moment of inertia of the disk.

$$I_{disk} = I_{total} - I_{particle} = (4mR^2 + \frac{3}{2}MR^2) - (m(2R)^2) = \frac{3}{2}MR^2 \rightarrow$$
$$I_{disk} = \frac{3}{2}MR^2$$

(Note that if students are familiar with the Parallel Axis Theorem this can be determined much more simply, though the PAT is not formally within the scope of the AP Physics 1 curriculum.)

14. A round object of mass m and radius r sits at the top of a track of length L inclined at an angle of θ with the horizontal.

(a) Derive an expression for the gravitational potential energy of the object in terms of m, L, and θ.

A student wishes to experimentally determine the object's moment of inertia, I, by adjusting the angle of the ramp and observing the behavior of the object as it rolls down the ramp without slipping.

(b) Describe an experimental procedure the student could use to collect the data necessary, including any equipment the student would need. Your description should include a listing of the independent and dependent variable(s).
(c) Derive an expression for the velocity of the object at the bottom of the ramp in terms of the length of the ramp, L, and the time it takes to travel down the ramp, t.
(d) Derive an expression for the moment of inertia of the object in terms of m, r, θ, t, L, and any fundamental constants.
(e) The student plots the square of the time it takes for the object to travel down the ramp, t^2, as a function of $1/\sin(\theta)$ to obtain a linear graph. How should the student determine the moment of inertia of the object from this graph? Highlight the calculation(s) used.
(f) Now suppose that you were not given the radius of the object. Describe an experimental procedure you could use to determine it, including any equipment you would need.
(g) Derive the expression for the moment of inertia of the object in terms of m, r, θ, t, L, and any fundamental constants using an alternate approach to your method from part (d).

Answer:

(a) $U_g = MgL\sin\theta$

(b) Using a stopwatch, the student could start the stopwatch when the object is released from rest, and halt the stopwatch when the object reaches the bottom of the incline, repeating this for varying levels of inclination and recording the independent variable (θ) and the dependent variable (t) in a data table. Other equivalent answers are acceptable, such as utilizing photogates to determine the time, or determining the speed of the object as it reaches the bottom of the incline.

(c) $v = \dfrac{2L}{t}$

(d) $mgL\sin\theta = \frac{1}{2}mv^2 + \frac{1}{2}I\omega^2 \rightarrow 2mgL\sin\theta = mv^2 + I\omega^2 \rightarrow I = \dfrac{2mgL\sin\theta}{\omega^2} - \dfrac{mv^2}{\omega^2}\xrightarrow[\omega=v/r]{v=\omega r}$

$I = \dfrac{2mgL\sin\theta r^2}{v^2} - mr^2 = mr^2\left(\dfrac{2gL\sin\theta}{v^2} - 1\right)\xrightarrow{v^2 = 4L^2/t^2} I = mr^2\left(\dfrac{2gLt^2\sin\theta}{4L^2} - 1\right)\rightarrow$

$I = mr^2\left(\dfrac{gt^2\sin\theta}{2L} - 1\right)$

(e) Plotting the function $t^2 = \left[\dfrac{2L(I + mr^2)}{mr^2 g}\right]\left(\dfrac{1}{\sin\theta}\right)$ indicates that the

slope of the line will give you the bracketed expression. Therefore, to find the moment of inertia, find the slope of the line first, then you may rearrange this equation to find $I = mr^2\left(\dfrac{g\times slope}{2L} - 1\right)$

(f) One possible answer includes wrapping a string around the circumference of the object, marking off the circumference, and measuring its length. The radius, then, is the circumference divided by 2π.

(g) The moment of inertia may be determined by a conservation of energy approach (per guidance above), as well as from a Newton's 2nd Law approach:

$F_{net_x} = mg\sin\theta - f = ma$

$\tau_{net} = fr = I\alpha \rightarrow f = \dfrac{I\alpha}{r}$

$mg\sin\theta - \dfrac{I\alpha}{r} = ma \xrightarrow{\alpha = a/r} mg\sin\theta - \dfrac{Ia}{r^2} = ma \rightarrow \dfrac{Ia}{r^2} = mg\sin\theta - ma \rightarrow$

$I = mr^2\left(\dfrac{g\sin\theta}{a} - 1\right)\xrightarrow{a = 2L/t^2} I = mr^2\left(\dfrac{gt^2\sin\theta}{2L} - 1\right)$

Gravity

1. A satellite of mass m_s orbits a planet of mass m_p at an altitude equal to twice the radius (R) of the planet. What is the satellite's speed assuming a perfectly circular orbit?

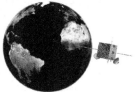

(A) $v = \sqrt{\dfrac{Gm_p}{R}}$

(C) $v = \sqrt{\dfrac{Gm_s}{2R}}$

(B) $v = \sqrt{\dfrac{Gm_s}{R}}$

(D) $v = \sqrt{\dfrac{Gm_p}{3R}}$

Answer: (D) $v = \sqrt{\dfrac{Gm_p}{3R}}$

Students must first recognize that the radius of the satellite's orbit is 3R, the radius of the planet plus the altitude of the satellite above the surface of the planet. Then, a force analysis recognizing the gravitational force of attraction provides a centripetal force yields:

$$F_{net_C} = F_g = ma_c = \frac{mv^2}{r} \rightarrow \frac{m_s v^2}{3R} = \frac{Gm_s m_p}{(3R)^2} \rightarrow v^2 = \frac{Gm_p}{3R} \rightarrow v = \sqrt{\frac{Gm_p}{3R}}$$

---

2. A spaceship in a circular orbit 400 km above the surface of the Earth wishes to manipulate its orbit to reach a point P on the opposite side of the Earth which is 1000 km above the Earth's surface. If the spaceship is at the position shown in the diagram and currently moving in a clockwise direction, in which direction should the ship accelerate in order to reach point P?

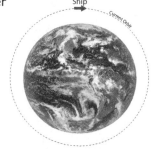

(A) toward the top of the page

(B) toward the right of the page

(C) toward the bottom of the page

(D) toward the left of the page

Answer: (B) toward the right of the page

To increase the radius of its orbit, the ship must attain a higher velocity, which requires an acceleration in the direction of its current velocity, or to the right of the page as depicted in this diagram. This will shift the orbit from a circular orbit to an elliptical orbit, and allow the ship to reach point P. (Note that a second acceleration will be required upon reaching point P if the ship wishes to maintain a circular orbit 1000 km above the Earth's surface.)

3. A rock is thrown horizontally from the top of a 100-meter-high vertical cliff on Planet Unicorn with a speed of 20 m/s. If the mass of Planet Unicorn is 10^{25} kg and the top of the cliff is approximately 4000 kilometers from the center of the planet, how far from the base of the cliff does the rock land?

(A) 0.022 m

(C) 43.8 m

(B) 0.044 m

(D) 90.1 m

Answer: (C) 43.8 m

The acceleration due to gravity is the gravitational field strength, which can be determined from Newton's Law of Universal Gravitation. The horizontal distance traveled by the rock is the time it takes for the rock to strike the ground (a kinematics exercise) multiplied by the horizontal velocity of the rock (given in the problem).

$$\Delta x = v_x t = (20 \, ^m\!/_s) \sqrt{\frac{2h}{g}} = (20 \, ^m\!/_s) \sqrt{\frac{2hr^2}{Gm}} = (20 \, ^m\!/_s) \sqrt{\frac{2(100m)(4000000m)^2}{6.67 \times 10^{-11} \, ^{N \bullet m^2}\!/_{kg^2}(10^{25} kg)}} = 43.8m$$

---

4. Which of the following changes would increase the magnitude of the gravitational field intensity an object feels when near a planet? (Select two answers.)

(A) increase the mass of the object

(B) increase the mass of the planet

(C) decrease the spin rate of the planet

(D) decrease the separation distance between object and planet

Answers: (B) and (D)

By definition field intensity, $g = F_g/m_o$ where m_o is the mass of the object. This equation expands to $g = Gm_p/r^2$ leading to choices (B) and (D)

---

5. Marty is an astronaut who is preparing to go on a mission in orbit around the Earth. For health reasons, his mass needs to be determined before take-off and while he is in orbit. The morning of the launch, Marty sits on one pan of a two-pan scale and 94 kg of mass is needed to balance him.

   (a) State and explain whether the two-pan scale registered Marty's gravitational mass or inertial mass.
   (b) After a few days in orbit Marty is again to determine his mass. Explain why the two-pan scale used before launch cannot be used to measure his mass while in orbit.
   (c) To determine Marty's mass in orbit he is to sit in a chair of negligible mass that is attached to a wall by a spring that has a force constant, k. Consequently, the chair freely vibrates back and forth with a period, T when displaced sideways a distance, x. Explain how the spring-mounted chair can be used to determine Marty's mass, m. Give relevant measurements and equation(s).
   (d) If Marty has lost mass while in orbit, what specific change would occur when he sits in the chair and starts it oscillating?
   (e) Explain why this spring-mounted chair measures Marty's inertial mass.

Answer:

   (a) Gravitational mass. The pans of the scale balance under the influence of gravity, not any other force.
   (b) In orbit the effects of gravity are not felt because everything is in free-fall. Consequently the scales will not become balanced or unbalanced when objects are placed on them.
   (c) Marty's mass can be determined by measuring the period of vibration of the oscillating chair (displacement is irrelevant) and using the equation ($T = 2\pi\sqrt{m/k}$).
   (d) The period of oscillation would decrease (no change in displacement).
   (e) Inertial mass affects an object's response to a non-gravitational force as described by Newton's 2nd Law of Motion. In this situation, the force is due to the spring and the response is the period of oscillation.

6. A space probe is sent on a mission to map out the gravitational field intensity in the vicinity of a satellite of planet X. Some of the data collected is shown in the chart below:

distance to satellite ($\times 10^6$ m)	field intensity (N/kg)
2.0	13.3
2.5	8.4
3.0	5.9
3.5	4.5
4.0	3.3
6.0	1.5

(a) On the axes below, plot the gravitational field intensity, g, vs. the distance, R, to the satellite.

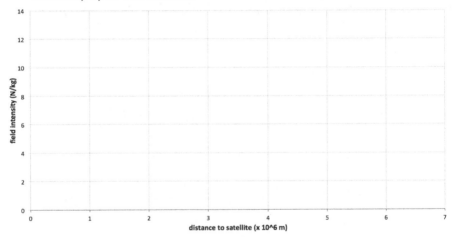

(b) Draw in the appropriate best fit line or curve.
(c) Using the best fit, what distance corresponds to a field intensity of 2.1 N/kg?
(d) In order to determine the mass of the satellite, a plot of field intensity vs. $1/R^2$ can be utilized. Fill in the appropriate values for $1/R^2$ in the chart below.

$1/R^2$ ($1/m^2$)	field intensity (N/kg)
	13.3
	8.4
	5.9
	4.5
	3.3
	1.5

(e) On the axes below plot gravitational field intensity, g, vs. $1/R^2$.

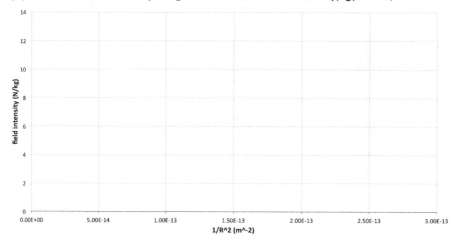

(f) Use the plot and best fit to determine the mass of the planet.

Answer:

(a)
(b)

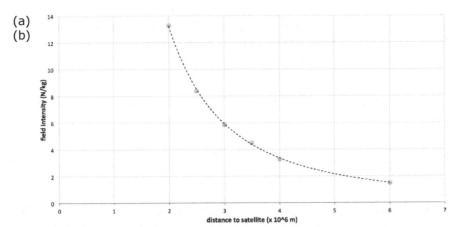

(c) distance of 5.0 ± .2×10⁶ meters (corresponds to g=2.1 N/kg)

(d)

1/R² (1/m²)	field intensity (N/kg)
2.5×10⁻¹³	13.3
1.6×10⁻¹³	8.4
1.1×10⁻¹³	5.9
8.2×10⁻¹⁴	4.5
6.3×10⁻¹⁴	3.3
2.8×10⁻¹⁴	1.5

(e)

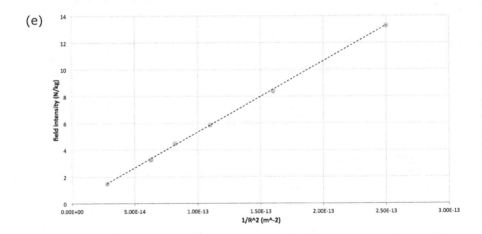

(f) slope = gR^2 = 5×10¹³ m³/s². If g = Gm_p/R^2 then m_p = $(gR^2)/G$ = slope/G = 7.5×10²³ kg

Oscillations

1. Which of the following statements about a spring-block oscillator in simple harmonic motion about its equilibrium point is false?

(A) The displacement is directly related to the acceleration.

(B) The acceleration and velocity vectors always point in the same direction.

(C) The acceleration vector is always toward the equilibrium point.

(D) The acceleration and displacement vectors always point in opposite directions.

Answer: (B) The acceleration and velocity vectors always point in the same direction. A, C, and D are all true, but the acceleration and velocity vectors sometimes point in the same direction, and sometimes point in opposite directions.

2. Which of the following are most likely to result in simple harmonic motion? Select two answers.

(A) A hole is drilled through one end of a meter stick, which is hung vertically from a frictionless axle. The bottom of the meter stick is displaced 12 degrees and released.

(B) A hole is drilled through one end of a meter stick, which is hung vertically from a rough axle. The bottom of the meter stick is displaced 12 degrees and released. Every time the meter stick swings back and forth the axle squeaks.

(C) A block is hung vertically from a linear spring. The opposite end of the spring is attached to a stationary point. The entire apparatus is placed in deep space. The block is displaced 4 cm from equilibrium and released.

(D) A block is placed on a frictionless surface and attached to a non-linear spring. The opposite end of the spring is attached to a wall. The block is displaced 2 cm from equilibrium and released.

Answers: (A) and (C). Simple harmonic motion requires a linear restoring force. The real pendulum described in answer (A) can be readily approximated using simple harmonic motion using the small angle approximation. Answer (B) can be eliminated due to the loss of energy from the rough axle, which will result in damping. Answer (C) describes true simple harmonic motion in a frictionless environment with the spring-block oscillator. Answer (D) can be eliminated due to the non-linear spring.

3. A spring of spring constant k is hung vertically from a fixed surface, and a block of mass M is attached to the bottom of the spring. The mass is released and the system is allowed to come to equilibrium as shown in the diagram at right.

(a) Derive an expression for the equilibrium position of the mass.
(b) The spring is now pulled downward and displaced an amount A. Derive an expression for the potential energy stored in the spring.
(c) At time t=0, the spring is released. Derive an expression for the period of the spring-block oscillator.
(d) Describe an experimental procedure you could use to verify your derivation of the period. Include all equipment required.
(e) How would you analyze the data to determine whether the experimental data verifies your derivation? What evidence from the analysis would be used to make the determination?

Answer:

(a) $y_{eq} = \dfrac{Mg}{k}$

(b) $U_s = \dfrac{1}{2}kA^2$

(c) $T_s = 2\pi\sqrt{\dfrac{M}{k}}$

(d) One method could involve utilizing a stopwatch to measure the time it takes for 10 oscillations, and dividing that by 10 to obtain the experimental period for the given mass. A more detailed analysis could include repeating this for a variety of masses. Equipment required would include a stopwatch and hanging masses. A variety of alternate acceptable answers exist, which could include, but are not limited to, electronic measuring devices such as photogates.
(e) One method would involve plotting the period vs. the square root of the mass. This plot should be linear, with the slope equal to 2π divided by √k. Including an uncertainty analysis of the data, if the calculated value for the spring constant from the slope of the graph matches the spring constant of the spring (within its uncertainty), it would be reasonable to conclude that the derivation is correct.

4. A spring of spring constant 40 N/m is attached to a fixed surface, and a block of mass 0.25 kg is attached to the end of the spring, sitting on a frictionless surface. The block is now displaced 15 centimeters and released at time t=0.

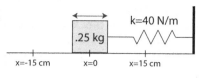

x=-15 cm   x=0   x=15 cm

(a) Draw a free body diagram for the block at t=0.
(b) Determine the period of the spring-block oscillator.
(c) Determine the speed of the block at position x=0.
(d) Write an expression for the displacement of the block as a function of time.
(e) Plot the displacement, speed, and acceleration of the mass as a function of time. Explicitly label axes with units as well as any in-tercepts, asymptotes, maxima, or minima with numerical values or algebraic expressions, as appropriate.
(f) Plot the kinetic energy of the mass and the elastic potential energy of the spring as functions of time. Label your plots K and U_s. Ex-plicitly label axes with units as well as any intercepts, asymptotes, maxima, or minima with numerical values or algebraic expressions, as appropriate.

Answer:

(a)

(b) $T_s = 2\pi\sqrt{\dfrac{M}{k}} = 0.497s$

(c) $\frac{1}{2}kx^2 = \frac{1}{2}mv^2 \rightarrow v = \sqrt{\dfrac{kx^2}{m}} = \sqrt{\dfrac{(40)(.15)^2}{.25}} = 1.90\,{}^{m}\!/_{s}$

(d) $x(t) = A\cos(\omega t) \rightarrow x(t) = 0.15\cos(12.6t)$

(e)

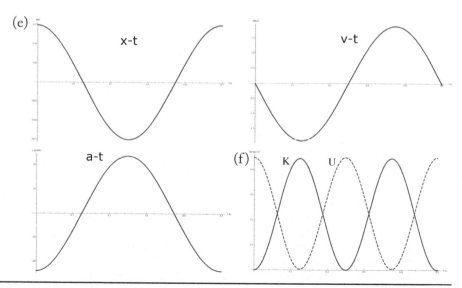

x-t

v-t

a-t

(f)   K   U

5. Students are to conduct an experiment to investigate the relationship between the length of a pendulum and its period. Their procedure involves hanging a 0.5 kg mass on a string of varying length, L, setting it into oscillation, and measuring the period. The students conduct the experiment and obtain the following data.

Trial	1	2	3	4	5	6
Length (m)	0.25	0.50	0.75	1.0	1.5	2.0
Period (s)	1.0	1.4	1.7	2.0	2.5	2.8

(a) Plot the period of the pendulum T as a function of the length of the string, L, and draw a best-fit curve. Label the axes as appropriate.
(b) Plot a linear graph as a function of L. Use the empty boxes in the data table to record any calculated values you are graphing. Label the axes as appropriate.

Students are then given strings of various lengths, a hanging mass, a meter stick, a stopwatch, appropriate survival equipment, and are transported to the surface of Planet X. There, they are asked to determine the acceleration due to gravity on the surface of Planet X using just this equipment.

(c) Describe an experimental procedure that the student could use to collect the necessary data as accurately as possible.

In order to determine the acceleration due to gravity, the students then create a plot of T vs. the square root of L, as shown below.

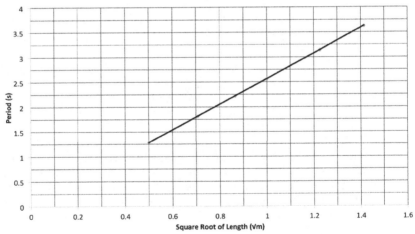

(d) Using the graph, calculate the acceleration due to gravity on Planet X.
(e) Following further analysis and experiments, a comparison of the students' experimental value for the acceleration due to gravity on Planet X is greater than the actual acceleration due to gravity on Planet X. Offer a reasonable explanation for this difference.

Answer:

(a)

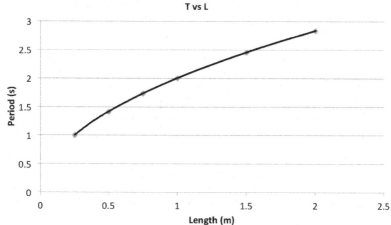

(b)

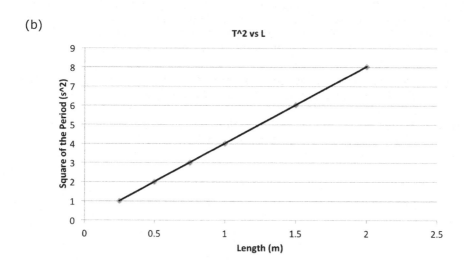

(c) Hang the mass from the string and attach the top of the string to a fixed point. Pull the mass back a small angular displacement ($\sim$ 10 degrees) and release. Use a stopwatch to time 20 oscillations, and divide the total time by 20 to obtain the period.

(d) $g = \dfrac{4\pi^2}{slope^2} = \dfrac{4\pi^2}{(2.5651)^2} = 6\,m/_{s^2}$

(e) Our analysis assumes the string is massless. A real string has mass, which would lead to a smaller measured period of oscillation, and therefore a larger measured acceleration due to gravity compared with the actual acceleration due to gravity on the planet.

8. A disk with a mass M, a radius R, and a rotational inertia of $I = \frac{1}{2} MR^2$ is attached to a horizontal spring which has a spring constant of k as shown in the diagram. When the spring is stretched by a distance x and then released from rest, the disk rolls without slipping while the spring is attached to the frictionless axle within the center of the disk.

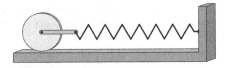

(a) Calculate the maximum translational velocity of the disk in terms of M,R, x, k.
(b) What would happen to the period of this motion if the spring constant of the spring increased? Justify your answer.
(c) What would happen to the period of this motion if the surface was now frictionless and the disk was not allowed to roll? Justify your answer.

Answer:

(a) Use conservation of energy to find the maximum velocity:

$$\tfrac{1}{2}kx^2 = \tfrac{1}{2}mv^2 + \tfrac{1}{2}I\omega^2 \xrightarrow[\substack{\omega=\frac{v}{R}}]{I=\frac{1}{2}MR^2} v_{max} = x\sqrt{\frac{2k}{3M}}$$

(b) If the spring constant k increased, the period of the motion would decrease. When you increase the k value of a spring the period of the spring decreases. This is due to the velocity of the object attached to the spring increasing. You can also look at the answer in part (a) and see that the velocity increases with a larger k which means the time to travel one full oscillation will decrease.
(c) There is a decrease in the period if the system is now on a frictionless surface. The disk will no longer roll and no energy is being put into rolling the disk. The total energy of the system will stay the same but more energy is now available for the translational kinetic energy (assuming we stretch the spring the same distance x). This leads to a higher velocity, which leads to a decrease in the period.

7. A 5-kg mass attached to a linear spring undergoes simple harmonic motion along a frictionless tabletop with an amplitude of 0.35 m and frequency of 0.67 Hz.

(a) Explain what the two criteria are for simple harmonic motion.
(b) Calculate the value of the spring constant.
(c) What is the ratio of the mass's acceleration when it is at half its amplitude to its acceleration when it is at full amplitude?
(d) Sketch a graph of the mass's acceleration as a function of its velocity for half of a cycle. Start when the mass is at its greatest positive amplitude. Mark the initial acceleration as a_0.

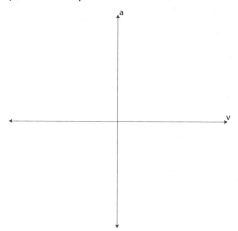

Answer:

(a) restoring force must be proportional to displacement; restoring force always opposes motion of mass

(b) $f = \dfrac{1}{2\pi}\sqrt{\dfrac{k}{m}} \rightarrow k = (2\pi f)^2 m = 89\,{}^{N}\!/\!_{m}$

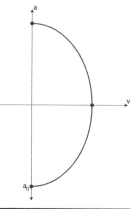

(c) The acceleration is proportional to the force which is proportional to the displacement according to Hooke's Law, therefore the ratio of a/a_0 is 0.5.

(d) When x is at a max, v is 0 and a is at a max. Acceleration is negative when displacement is positive. Velocity is positive for the entire half cycle. Velocity is proportional to the sine function and acceleration is proportional to the cosine function (out of phase), resulting in an elliptical plot (full credit for the three marked points regardless of shape of the graph).

8. On a flat surface in Toledo, Ohio, a simple pendulum of length 0.58 m and mass 0.34 kg is pulled back an angle of 45° and released.

(a) Determine the theoretical period of the pendulum.
(b) When the period is measured (with a photogate, for 20 oscillations) it is found that the experimental period is 5 percent higher than expected. Repeated measurements consistently give similar values. What is the most likely explanation for this systematic error?
(c) What would be the period of this pendulum if it was in a free-fall environment? Explain.

Answer:

(a) $T_p = 2\pi\sqrt{\dfrac{l}{g}} = 2\pi\sqrt{\dfrac{0.58m}{9.8\,m/_{s^2}}} = 1.53s$

(b) The most likely explanation is the theoretical equation is under-calculating the period. The equation used in part (a) is based on the small angle approximation and in the lab situation the angle is 45°, which is too large an angle to apply the small angle approximation. Consequently a large angle gives a slightly greater period than expected.
(c) There would be no period (or T = ∞) because the effects of gravity are not felt in free-fall.

9.) A group of students want to design an experiment where they test the period of simple pendulum as a function of the acceleration due to gravity, g. Part of their motivation is to verify the equation of the period of a simple pendulum:

$$T_p = 2\pi\sqrt{\frac{l}{g}}$$

They propose to bring their apparatus on NASA's Zero-g airplane that can create a range of values of g for short periods of time. Their proposal calls for three trials each for two situations where g < 9.8 m/s² (where everything feels lighter than normal), g = 9.8 m/s² and two situations where g > 9.8 m/s² (where everything feels heavier than normal).

(a) If the Zero-g airplane moves in a continuous series of up-and-down parabolas, as shown at right, identify at which position(s) in its trajectory someone would feel heavier than normal.

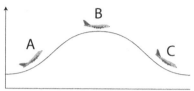

(b) If the simple pendulum the students bring consists of a 1.0 kg sphere with a string attached, discuss two other important factors that must be controlled throughout the experiment and how they will be controlled.

(c) Since the students have a limited number of trials for each value of g, they need a method of minimizing uncertainty in the measurement of the period of the pendulum. Discuss a method for doing so. Be sure to list necessary equipment and how it will be applied.

(d) Assuming that the students carry out the experiment and gather valid data, they want to graph their data. If they plot the variable T_p^2 on the y-axis, what variable should they plot on the x-axis so they get a straight line best fit?

(e) When they draw the line of best fit, the students calculate a slope of 9.8 m. What is the length of their pendulum?

Answer:

(a) Heavier than normal at positions A and C.
(b) Length of string can be controlled by securing one end to the sphere and the other one to a clamp system. The angle (amplitude) can be controlled by pulling back a predetermined (small) distance. Perhaps have a barrier to pull back to each time.
(c) Let the pendulum swing back and forth 5 to 10 times and time with a stopwatch or photogate system.
(d) Plot T^2 vs 1/g (accept T^2 vs. k/g where k is some constant value such as $4\pi^2 L$ or L ...)
(e) Assuming they plotted T^2 vs 1/g, the slope is proportional to T^2g.
Using the pendulum equation L = $(T^2g)/ 4\pi^2$ = slope/$4\pi^2$ = 0.25 m

Mechanical Waves

1. A fire truck is moving at a fairly high speed, with its siren emitting sound at a specific pitch. As the fire truck recedes from you which of the following characteristics of the sound wave from the siren will have a smaller measured value for you than for a fireman in the truck? Choose two characteristics.

(A) frequency

(C) speed

(B) wavelength

(D) intensity

Answer: (A) and (D) frequency and intensity.

Doppler effect shifts frequency to a lower measured value as fire truck recedes and a longer wavelength. There is no change in the sound's speed. Intensity is lower for you because the siren is further away from you than the fireman.

2. A student tunes her guitar by striking a 110-Hertz A-note on a tuning fork, and simultaneously playing the 5th string on her guitar. Listening closely, she hears the amplitude of the combined sound oscillating twice per second. Which of the following is most likely the current frequency of the 5th string on her guitar?

(A) 108 Hertz

(B) 114 Hertz

(C) 220 Hertz

(D) 440 Hertz

Answer: (A) 108 Hertz

The beat frequency is the difference in frequency between the two waves.

3. A transverse wave travels in medium X with a speed of 800 m/s and a wavelength of 4 m. The wave then moves into medium Y, traveling with a speed of 1600 m/s.

  (a) Determine the frequency of the wave in medium Y.
  (b) Determine the wavelength of the wave in medium Y.

Answer:

  (a) 200 Hz (frequency does not change when a wave enters a new medium).

  (b) $\lambda = \dfrac{v}{f} = \dfrac{1600\,m/s}{200\,Hz} = 8m$

---

4. Wave pulses travel toward each other along a string as shown below. Answer the following questions in terms of the resulting superposition of the pulses when their centers are aligned.

A   B   C   D

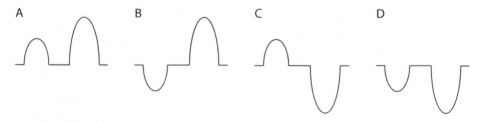

(A) Rank the maximum amplitude of the resulting superposition from smallest to largest.

(B) Rank the magnitude of the maximum amplitude of the resulting superposition from smallest to largest.

Answer: (A) D, C, B, A   (B) B=C, A=D

5. A string on a musical instrument is fixed at both ends. If the length of the string is 0.3 meters and waves travel through the string with a speed of 450 m/s, which of the following frequencies would you expect to hear from the string? Select two answers.

(A) 570 Hz

(B) 750 Hz

(C) 1125 Hz

(D) 1500 Hz

Answer: (B) 750 Hz and (D) 1500 Hz

The frequencies produced by waves on a string are given by f=nv/(2L).

---

6. A baseball is moving to the right with a speed of v. At four different positions people have radar guns pointed at the ball to measure a Doppler shift in frequency in order to determine the baseball's speed, as shown in the diagram below right (note that wave fronts are NOT drawn to scale). Rank the measured shift in frequency of the radar beam from lowest to highest based on the position of the radar gun.

(A) 3 < 4 < 2 < 1

(B) 3 < 1 < 2 < 4

(C) 1 < 2 < 4 < 3

(D) 4 < 2 < 3 < 1

Answer: (D) 4 < 2 < 3 < 1

The amount of shift depends on relative velocity in the line of motion of the baseball. Position 1 is directly in line, whereas position 4 has no component of the velocity coming at it. Position 3 has a relatively large component and position 2 a small component.

7. Sound waves are traveling through air when they encounter a steel barrier. Some of the sound waves are reflected and inverted, while the rest are transmitted through the steel barrier. The restoring forces within the steel are significantly higher than that of air. Which of the following changes occur to the sound waves at the air/steel boundary? Select two answers.

(A) their amplitude decreases in the steel because some of the waves are reflected.

(B) their speed increases in the steel because the restoring forces are higher.

(C) their frequency decreases in the air because the restoring forces are lower.

(D) their wavelength decreases in the air because the reflected waves are inverted.

Answer: (A) and (B)

(A) true b/c transmitted wave has less energy

(B) true b/c speed is proportional to the square root of the restoring force

(C) false b/c frequency does not change at all

(D) false b/c wavelength doesn't depend on phase

8. Students are attempting to determine the speed of sound in air using tuning forks and tubes which are closed at one end. In this procedure, the tube is filled with water, and a tuning fork of known frequency is struck. The vibrating tuning fork is then held over the tube filled with water, and the water is slowly drained out of the tube while students listen for the loudest possible sound at the first resonant condition. Once the loudest possible sound is heard (the first harmonic), the distance from the top of the tube to the water's surface (L) is measured and recorded. This procedure is repeated for five tuning forks of varying frequencies. Data is recorded in the table below.

Trial	1	2	3	4	5	6
Freq (Hz)	128	256	288	384	426.6	512
Period (s)						
L (m)	.65	.33	.29	.23	.19	.17
λ (m)						

  (a) Determine the period of oscillation (T) for each of the five trials and fill in the data table above.
  (b) Write an equation for the wavelength of the sound wave (λ) as a function of L.
  (c) Use the grid below to plot a linear graph of wavelength (λ) as a function of period (T). Use the empty boxes in the data table to record any calculated values you are graphing. Label the axes as appropriate.

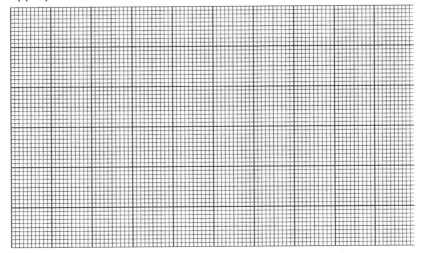

  (d) Draw a best-fit line on your graph. Using your best-fit line, determine the speed of the sound waves in air.
  (e) Describe how your procedure and analysis would change if you used the third harmonic instead of the first harmonic to determine the speed of sound. Indicate specifically any changes in calculations.

Answer:

(a) Data is recorded in the table below.

Trial	1	2	3	4	5	6
Freq (Hz)	128	256	288	384	426.6	512
Period (s)	.00781	.00391	.00347	.00260	.00234	.00195
L (m)	0.65	0.33	0.29	0.23	0.19	0.17
λ (m)	2.6	1.32	1.16	0.92	0.76	0.68

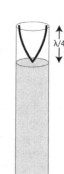

(b) Equation: λ=4L

(c)

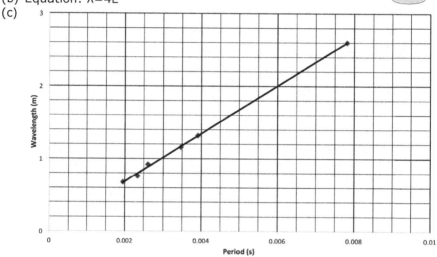

(d) slope=v=335 m/s

(e) In utilizing the third harmonic, you would continue to drain the water from the tube until the second time you heard the increase in amplitude. Then, you would determine the wavelength using λ=4L/3.

Electricity

1. A student wearing shoes stands on a tile floor. The students shoes do not fall into the tile floor due to

(A) a force of repulsion between the shoes and the floor due to macroscopic gravitational forces.

(B) the electrical repulsion between the shoes and the floor due to inter-atomic electric forces.

(C) the strong nuclear force creating a force of repulsion between the atoms in the student's shoes and the atoms in the tile.

(D) The weak nuclear force creating a force of repulsion between the leptons in the student's shoes and the nucleons in the tile.

Answer: (B) the electrical repulsion between the shoes and the floor due to interatomic electric forces.

Contact forces result from the interaction of one object touching another object and they arise from interatomic electric forces.

---

2. A neutral baryon is compared of three quarks: two down quarks which each have a charge of -⅓e, where e is the elementary charge, and an up quark. What is the charge on an up quark?

(A) +⅓e

(B) -⅓e

(C) +⅔e

(D) -⅔e

Answer: (C) +⅔e

The entire baryon is neutral, so the sum of charges from its constituent particles must be zero due to the law of conservation of charge. Therefore, the charge on the up quark must be (C) +⅔e.

3. Following a period of extended use, the battery-powered circuit at right no longer functions. What are reasonable explanations for the circuit no longer functioning? Select the two best answers.

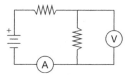

(A) The electrons from the battery are used up as the circuit "runs out of juice."

(B) The potential difference supplied by the battery has degraded.

(C) The current in the circuit has been converted to heat in the resistors.

(D) The battery can no longer supply sufficient power to operate the circuit.

Answers: (B) and (D)

Circuits do not run out of juice as electrons are used up -- this would violate the law of conservation of charge, so A cannot be an answer. The current in the circuit is not converted to heat, as current is a flow of electrons, and the law of conservation of charge prohibits the destruction of that charge, therefore C cannot be an answer. Both B and D are correct.

---

4. The magnitude of the force of attraction between two point charges is F. If the magnitude of each point charge is doubled, and the distance between the charges is doubled, what is the new magnitude of the force of attraction between the point charges?

(A) F/2

(B) F

(C) 2F

(D) 4F

Answer: (B) F

Doubling each charge doubles the force, for a total force of 4F. Then, doubling the distance cuts the force by a factor of 4 due to the inverse square nature of the charge separation, bringing the net force back down to its original value F.

5. Consider two circuits, A & B, shown below. Both have identical resistors of resistance R and identical light bulbs.

(A) Lightbulb Battery R R

(B) Lightbulb Battery R R

Correct statements about the bulbs in each circuit include which of the following? Choose 2 statements.

(A) Bulb A and Bulb B are the same brightness because the resistances are all the same.

(B) Bulb B has a greater voltage drop across it than Bulb A.

(C) The currents through each bulb are the same as the currents through their respective batteries.

(D) Bulb A has a greater current through it than Bulb B.

Answer: (B) and (C)

(A) is incorrect because the equivalent resistances of the circuits are different because of the differing series/parallel resistor configurations.

(B) is correct because the resistance of the resistor network in circuit B is less than that of A because the resistors are in series, leaving more of the battery's voltage to be dropped across the lightbulb in circuit B.

(C) is correct because in both cases, the current through the battery also flows through the light bulb due to conservation of charge / Kirchhoff's Current Law.

(D) is incorrect because the effective resistance of circuit A is greater than circuit B, therefore there is less current through Bulb A.

6. A scientist claims that she has captured and isolated a top quark and measured the magnitude of its charge as 5.33×10^{-20} C. Which of the following best explain why this claim should be doubted? Select two answers.

(A) The magnitude of the charge on a top quark is two-thirds of an elementary charge, which is twice as large as the charge the scientist measured.

(B) The smallest observed unit of charge that can be isolated is the elementary charge.

(C) Particles cannot have a charge less than an elementary charge.

(D) It is physically impossible to measure any charge smaller than an elementary charge.

Answer: (A) and (B) are the correct answers.

(A) the magnitude of the charge on a top quark is ⅔ of an elementary charge.

(B) The smallest observed unit of charge that can be isolated is the elementary charge.

(C) Is incorrect because particles can have a charge less than an elementary charge, but they cannot be isolated.

(D) Is incorrect because charges smaller than an elementary charge can be measured.

7. The circuit schematic at right shows four resistors and four ammeters connected to a voltage source, V. It is known that the current in ammeter A1 is 6.0 A, and the current in ammeter A2 is 2.0 A.

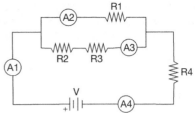

I. Which of the following must be true statements about the current through resistors R2 & R3?

(A) The current through both must be 4.0 A because they have the same voltage drop across them.

(B) The current through both must be 4.0 A because they are in series with each other.

(C) The current through both must be 2.0 A because they share the current in the lower branch.

(D) The current cannot be determined without information about the resistances of R2 & R3.

II. If the resistance of R4 is given, what other information is necessary to determine the value of V, the voltage of the battery?

(A) the reading of ammeter A4 and resistance of R2.

(B) the reading of ammeter A4 and resistance of R1.

(C) resistance of R1.

(D) resistances of R1, R2 & R3.

I. Answer: (B)

(A) Incorrect because one cannot assume the resistances (and therefore the voltage drops) are the same.

(B) Correct because the lower branch must draw 4A to conserve charge and the relative currents are the same irrespective of resistance values.

(C) Incorrect because there is no sharing of current in series.

(D) Incorrect - see (b) above.

II. Answer: (C)

(A) Incorrect because R2 & R4 do not complete a path in which the loop rule can be applied.

(B) Incorrect because even though R1 & R4 complete a loop, A4 is superfluous since it's in series with A1.

(C) Correct because R1 & R4 complete a loop and $V = I1R1 + I4R4$, where I1 is known from A1.

(D) Incorrect because knowledge of R1 (or R2 & R3) is all that is necessary, not both.

8. In the circuit shown below right, resistors R1, R2, R3, and R4 all have the same value of resistance. Removing which resistor from the circuit would result in the smallest change to the ammeter reading? Select two answers.

(A) R1

(B) R2

(C) R3

(D) R4

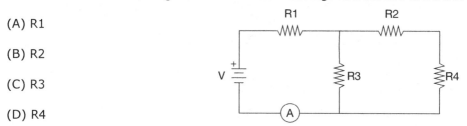

Answers: (B) R2 and (D) R4

The initial current flow in the circuit is 0.75 V/R. Removing either R2 or R4 from the circuit (or both) effectively makes the circuit a series circuit with just R1 and R3, for a current flow of 0.5 V/R, a reduction of 0.25 V/R. Removing R1 from the circuit opens the circuit and no current flows, and removing R3 creates a series circuit with an equivalent resistance of 3R, for a total current flow of 0.33 V/R. Therefore, removing R2 and/or R4 results in a circuit with the least change in current flow through the ammeter.

---

9. A +4 C point charge and a +2 C point charge are situated on an axis as shown in the diagram. In which region can you place an unknown point charge and have it remain at rest?

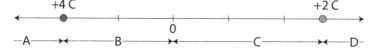

Answer: C

In order for the point charge to remain at rest, the net force on the charge must be zero, therefore the the force due to the +4 C charge must be equal in magnitude and opposite in direction to the force due to the +2 C charge. In regions A and D, the forces due to the charges are in the same direction, so those regions can be eliminated. In order for the magnitudes of the forces to be equal, the point must be closer to the +2 C charge than the +4 C charge, therefore region C must be the correct answer.

10. The graph at right shows a plot of voltage versus current for a filament lamp. The 12 V dashed line shows the voltage setting of the power source.

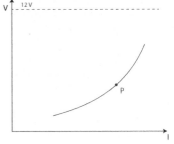

(a) Explain how the graph shows that the filament does not obey Ohm's Law.
(b) Give a physical explanation of why the filament lamp does not obey Ohm's Law.
(c) Explain how one would determine the resistance of the filament for point P on the graph.
(d) In order to gather the data to make the graph shown above, a variable resistor was placed in series with the filament lamp, and both were connected to a 12-volt source as shown in the schematic below.

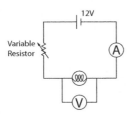

Explain why using a variable resistor in series with the lamp
i. will not allow a reading of 0 V across the filament lamp.
ii. will not allow a reading of 12 V across the filament lamp.

Answer:

(a) Ohm's law states that the relationship b/w V & I is linear and proportional. The graph shows a non-linear plot, thus does not obey Ohm's law;
(b) As the current increases the temperature of the filament increases. As the temperature increases the resistance increases.
(c) The resistance is found by taking the ratio of voltage to current at point P.
(d)
    i. To get a reading of 0 V across the bulb, the variable resistor would have to have an infinite resistance. An infinite resistance in series with the bulb would leave 0 V across the rest of the circuit.
    ii. To get a reading of 12 V across the filament lamp there would have to be 0 resistance in the variable resistor, the wires and the voltage source; this is not physically possible.

11. The circuit at right shows a light bulb of resistance 2R connected to a battery with no internal resistance and two resistors, each with a resistance R.

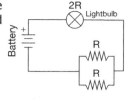

(a) Which component has more current passing through it, the light bulb or either of the resistors? Give a detailed explanation with supporting equations if necessary.

(b) Develop an expression for the voltage drop across the light bulb as a fraction of the battery voltage, V. Give detailed support.

(c) Suppose one of the resistors, R is removed from the circuit, without reconnecting the wires. Which component will develop more power, the remaining resistor or the light bulb? Justify your answer.

Answer:

(a) The light bulb has more current than either bulb because the bulb is in series with the battery, while the resistors are in parallel. All the current passes through the bulb and it splits equally between both resistors.

(b) 4/5 V. The total current in the circuit will be 2V/5R, therefore the potential drop across the bulb, by ohm's law, is equal to the product of the current through the bulb and the bulb's resistance, or $V_{bulb} = (2V/5R) \times (2R) = 4/5$ V.

(c) The light bulb will develop more power than the resistor. Removing on resistor will put the remaining resistor in series with the bulb, so both the bulb and the resistor receive the same current. The bulb, however, has twice the resistance of the resistor, which will therefore develop more power since $P = I^2R$.

You can find these problems, and more, in worksheet form, ready to print, directly from the APlusPhysics.com site at http://www.aplusphysics.com/ap1/ap1-supp.html

Index

terminal velocity 93
ticker-tape diagrams. See particle diagrams
topographic maps 290
torque 185
translational inertia 188
trigonometry 13–15, 75
troughs 239
tuning fork 243

U

uniform circular motion 164
unit analysis 107, 117
unit conversions 11–12
universal gravitation. See Newton's Law of Universal Gravitation
universal gravitational constant 202, 283

V

vector field 205
vectors 15–16, 50, 74, 165, 276
  components of 16–21, 56, 59, 75, 96–100, 106
  equilibrant of 21
velocity 24, 27–30, 37–40, 129
  recoil 151
velocity-time graph 36–37
VIRP table 306–310, 311–316
voltage. See also electric potential difference
voltage vs. current graph 298
voltaic cells 299
voltmeters 302
volts 284, 286
v-t graph. See velocity-time graph

W

watts 112, 301
wave equation 245–247
wavelength 239, 243
waves
  compressions 243–244
  crests 239, 251
  electromagnetic 238, 245
  interference 250–252
  longitudinal 238, 243, 247
  mechanical 238
  phase 239, 252
  radio 238
  rarefactions 243–244
  resonance 248
  seismic 238
  sound 240, 247–250
  standing 255–257
  transverse 238, 240
  troughs 239, 251
  velocity 245–247
  X-rays 238
weak nuclear force 268
weight 75–76. See also force: gravitational

weightlessness 175, 211–215
wires 295
work 3, 24, 106–109, 116, 117, 125, 285, 290, 300–301
Work-Energy Theorem 125–137